SpringerBriefs in Biology

For further volumes:
http://www.springer.com/series/10121

Naoyuki Fuse • Tasuku Kitamura
Takashi Haramura • Kentaro Arikawa
Michio Imafuku

Evolution in the Dark

Adaptation of *Drosophila* in the Laboratory

 Springer

Naoyuki Fuse
Global COE Program
Graduate School of Science
Kyoto University
Kitashirakawa-Oiwake, Sakyo-ku
Kyoto 606-8502, Japan

Tasuku Kitamura
Department of Zoology
Graduate School of Science
Kyoto University
Kitashirakawa-Oiwake, Sakyo-ku
Kyoto 606-8502, Japan

Takashi Haramura
Department of Zoology
Graduate School of Science
Kyoto University
Kitashirakawa-Oiwake, Sakyo-ku
Kyoto 606-8502, Japan

Kentaro Arikawa
Laboratory of Neuroethology
The Graduate University
 for Advanced Studies
Shonan Village, Hayama
Kanagawa 240-0193, Japan

Michio Imafuku
Department of Zoology
Graduate School of Science
Kyoto University
Kitashirakawa-Oiwake, Sakyo-ku
Kyoto 606-8502, Japan

ISSN 2192-2179   ISSN 2192-2187 (electronic)
ISBN 978-4-431-54146-2 ISBN 978-4-431-54147-9 (eBook)
DOI 10.1007/978-4-431-54147-9
Springer Tokyo Heidelberg New York Dordrecht London

Library of Congress Control Number: 2013948398

Printed on acid-free paper

Springer is part of Springer Science+Business Media (www.springer.com)

Foreword

While the twentieth century was the century when researchers tried to discover "the general basic principles of organisms," the twenty-first century is expected to be the century when researchers try to understand "the evolution and diversity of organisms" on the basis of such general principles of organisms by integrating various disciplines such as morphology, physiology, and ecology.

The chief difficulty in studying "the evolution and diversity of organisms" lies in the fact that we have to consider factors at various levels ranging from the genome to the ecosystem. As taking various factors into account may cause a loss of focus, traditional studies have been restricted to analyzing only one individual level or factor. However, unfortunately, the current research and education system based on such a compartmentalized approach is inadequate for incisively studying "the evolution and diversity of organisms."

In order to solve these problems, we should strongly emphasize the necessity for joint studies and integration of the education programs between micro-level biology (genomic science, evolutionary developmental biology, genetic science, cell biology, neurobiology, molecular physiology, and molecular evolutionary studies) and macro-level biology (primatology, anthropology, ethology, environmental biology, evolutionary taxonomy, and so on) to young biologists. We launched a new education program in Kyoto University, called "Global COE program for Evolution and Biodiversity Research" to promote such integrative studies at various levels, and have succeeded in initiating novel currents of study of biodiversity that led rather than followed those in the rest of the world. To this aim, we decided to publish six books in "SpringerBriefs in Biology" which we hope will stimulate interest in such novel approaches on the evolution and diversity of organisms in the world and among young biologists.

In this book, we focus on a fascinating case of sensing and the evolution of animals. What happens when fruit flies are kept in darkness for 58 years? A simple and very long project was started at Kyoto University in 1954: The "Dark-fly" has evolved in a dark environment for 1,400 generations. Is the Dark-fly adapted to a dark environment? What is the molecular nature of their adaptation? We describe here intriguing alterations of morphology and behaviors in the Dark-fly. Now, recent progress in genome science has enabled us to explore the genome of the Dark-fly. We describe adaptive traits and genome features of the Dark-fly, including mutations in genes encoding olfactory receptors, enzymes involved in eye pigmentation and xenobiotics, and a light receptor. The behavioral and physiological traits related to these mutations might be advantageous for life in a dark environment.

Kiyokazu Agata
Professor, Department of Biophysics, Kyoto University
Project Leader of Kyoto University Global COE program
"Evolution and Biodiversity"

Preface

Biological organisms are often remarkably adapted to diverse environments by specialized metabolisms, morphologies or behaviors. How organisms come to possess adaptive traits is a fundamental question for evolutionary biology. As one answer, Charles Darwin proposed the natural selection theory: adaptive traits are selected and consequently become prevalent in a population during its history. This concept is widely accepted, but its molecular mechanism is not fully understood. To address this issue, we utilize an unusual organism, the "Dark-fly."

In 1954, the late Dr. Shuiti Mori (Kyoto University) asked the question how the daily light conditions affect the properties of organisms. He therefore started to maintain a culture of the fruit fly, *Drosophila melanogaster*, in complete darkness. After his death, members of the laboratory inherited his project and have maintained this fly strain until today. As of 2013, "Dark-fly" has evolved in the dark for 58 years (1,400 generations). Relative to human history, this would be as if Cro-Magnon man had started living in the dark 28,000 years ago and had continued it until now. Imagine what might have happened to him. What has happened to Dark-fly in its 58-year history?

In this book, we describe the past and present studies of Dark-fly performed at Kyoto University. Firstly, we summarize the history of the Dark-fly project (Chap. 1 written by Tasuku Kitamura, Takashi Haramura, and Michio Imafuku). In the early phase of the project, Mori and others found that Dark-fly did not have poor eyesight, but rather exhibited higher phototactic ability compared with the light-reared control fly. Although Dark-fly displayed lengthened bristles on the head, its overall morphology looks normal and there is no apparent morphological feature related to the dark-adaptation. Instead, there is a considerable possibility that Dark-fly might have changed its behavior, physiology and/or metabolism for the dark-adaption.

Circadian rhythm is a key aspect of organisms under diurnal periodicity. Organisms adjust their behavior, physiology and metabolism to accord with alternating light and dark cycles. An intriguing issue is whether Dark-fly retains circadian rhythms (Chap. 2 written by Takashi Haramura and Michio Imafuku). In fact, although periodicity of locomotor activity is retained by Dark-fly, its rhythmic changes of activity seem to be weakened. Thus, in contrast to cave-dwelling organisms that have totally abolished circadian rhythms, Dark-fly is in the process of changing its circadian behavior.

Flies sense light mainly via photoreceptors in their compound eyes. Photoreceptors develop specialized cellular structures and complex neural circuits to sense light effectively and to discern objects precisely. Dark-fly develops normal structures of photoreceptors (Chap. 3 written by Michio Imafuku and Kentaro Arikawa) if it is reared with a nutrient-rich medium. In contrast, rearing with a nutrient-poor medium, which has been used for the maintenance of Dark-fly, results in the degeneration of photoreceptors for either Dark-fly or the control fly. Thus, diet significantly impacts photoreceptor development, but it is unclear how the nutrient conditions may have influenced the evolution of Dark-fly.

Reproductive success is one of the measures of adaptation. Dark-fly produces more offspring in the dark than in the light for a certain period, but the normal fly does not show such an advantage in the dark (Chap. 4 written by Naoyuki Fuse). Furthermore, Dark-fly females survive longer than the normal fly females, especially in the dark. These traits of Dark-fly would contribute to reproductive success that would potentially be adaptive for living in the dark.

Recent progress in genome science, especially as represented by next-generation sequencing, enabled us to perform whole-genome sequencing for Dark-fly (Chap. 4 written by Naoyuki Fuse). We thereby identified a large number of mutations in its genome. These include mutations in genes encoding a light receptor, olfactory receptors, and enzymes involved in detoxification and neural development. We thus obtained a list of candidate genes potentially involved in Dark-fly's traits. Functional analyses of the mutations remain as future issues.

The Dark-fly project is a simple but long-term experiment, and is proceeding even now. This remarkable project surely owes much to the freedom of ideas that is a treasured academic tradition of Kyoto University. Recently, the application of modern genome science to the project has enabled us to move toward understanding the molecular basis of Dark-fly's adaptation processes. We believe that this new phase of the project will strongly enhance the value of the Dark-fly for the field of evolutionary biology in the future.

Kyoto, Japan   *Naoyuki Fuse*

Acknowledgements

We thank Dr. Noriyuki Sato and Dr. Kiyokazu Agata for encouraging us to push the "Dark-fly" project into the new phase. We are grateful to the many people who have maintained the Dark-fly culture, and to Mr. Kohei Okamoto, Ms. Minako Izutsu, Mr. Yuzo Sugiyama, and Mr. Osamu Nishimura for performing the experiments and analyses described here.

We thank Dr. Akira Nagatani, Dr. Yoshinori Shichida, and Dr. Jun Zhou (Harvard University) for their helpful comments, and Dr. Hideharu Numata, Dr. Akira Mori, and many colleagues for their support and discussions. We are grateful to Dr. Masatoshi Tomaru (Kyoto Institute of Technology) and Dr. Kevin Cook (Indiana University) for supplying experimental flies and stock information, and to Dr. Asao Fujiyama (National Institute of Genetics) for genome sequencing.

We heartily thank Dr. Elizabeth Nakajima for carefully reading the manuscript. This work was supported in part by Global COE Program for Biodiversity and Evolution, Kyoto University.

Contents

1 History of the "Dark-fly" Project 1
  1.1 Introduction 1
  1.2 Details of the Maintenance of Dark-fly 2
  1.3 Summary of Experiments Performed So Far 4
      1.3.1 Response to Light 4
      1.3.2 Rhythm of Daily Emergence 8
      1.3.3 Olfactory Response 9
      1.3.4 Elongation of Head Bristles 10
      1.3.5 Fecundity 10
      1.3.6 Developmental Rate 10
  1.4 Reflections on the Dark-fly Project 12
  References 13

2 Circadian Rhythm of Dark-fly 15
  2.1 Introduction 15
  2.2 Materials and Methods 16
      2.2.1 Flies 16
      2.2.2 Recording of Activity 16
      2.2.3 Data Analysis 17
  2.3 Results 18
      2.3.1 Strength of Rhythmicity 18
      2.3.2 Entraining Ability 19
      2.3.3 Activity Pattern 20
  2.4 Discussion 20
  References 21

3 Compound Eyes of Dark-fly 23
  3.1 Introduction 23
  3.2 Materials and Methods 24
       3.2.1 Animals... 24
       3.2.2 Electron Microscopy.. 24
  3.3 Results and Discussion 24
  References.. 27

4 Genome Features of Dark-fly 29
  4.1 Introduction 29
       4.1.1 Dark-fly as a Model Organism for Studying Environmental
              Adaptation 29
       4.1.2 Organisms Living in Darkness 30
       4.1.3 *Drosophila* Research Using NGS Technology 33
  4.2 Traits of Dark-fly.. 34
       4.2.1 Fecundity of Dark-fly 34
       4.2.2 Longevity of Dark-fly... 37
  4.3 Genome of Dark-fly... 39
       4.3.1 Whole Genome Sequencing for Dark-fly 39
       4.3.2 Analysis of Dark-fly SNPs and InDels 43
       4.3.3 Nonsense Mutations in the Dark-fly Genome 44
       4.3.4 Genome Regions Selected in Dark-fly 46
       4.3.5 SNPs and InDels in the Selected Genome Regions.................. 49
  4.4 Conclusions and Future of Dark-fly 51
  References.. 52

Chapter 1
History of the "Dark-fly" Project

Abstract We summarize the history of the "Dark-fly" project here. We describe the method of maintenance and results of experiments performed by Shuiti Mori, who initially started this project about 60 years ago. We also conducted statistical re-examination of some data. As a result, we concluded that differences between the "Dark-fly" and a control fly were apparent in response to light, olfactory sense, and length of sensory hairs. Finally, we discuss future prospects of this project.

Keywords Dark environment • *Drosophila melanogaster* • Olfactory sense • Photokinesis • Phototaxis • Sensory hairs

1.1 Introduction

One of the characteristics that all organisms show is adaptation to changes in the environment where they live. About 60 years ago, the late Professor Shuiti Mori of Kyoto University and his colleagues initiated a series of experiments that addressed how the characteristics of fruit flies changed in accordance with changes in the environment. One of their experiments, for example, showed that when fruit flies were exposed to the culture medium containing a toxic substance, copper sulfate, their progeny reared with the normal non-toxic medium displayed increased resistance to this substance in comparison with control flies that had been reared with the normal medium (Osawa et al. 1958). They also investigated the effects of types of culture media on fertility (Mori and Yanagishima 1957; Mori 1957a, b), and of methylene blue on oxygen consumption (Mori and Yanagishima 1954).

As a part of that series of experiments, the "Dark-fly" project was devised, namely, some lines of the fruit fly *Drosophila melanogaster*, Oregon-R-S strain, were maintained in complete darkness over many generations to explore any changes of characters that were related to long-term life in the dark. At present, these "Dark-flies" have passed more than 1,400 generations in the dark (Figs. 1.1 and 1.2). During this period, many experiments have been performed on these flies, and ever

Fig. 1.1 Morphology of Dark-fly. Dark-fly (*right*) looks similar to normal fly (*left*). Dark-fly possesses eyes and pigmented cuticles and does not show apparent morphological traits related to the adaptation

Fig. 1.2 Notebooks recording details of all transferring of the Dark-fly strain through June 2012

now new findings are being accumulated on these flies, as described in other chapters of this book. Here, we describe the details of the maintenance of the Dark-fly, summarize the results of some experiments performed so far, and present a general discussion of this project.

1.2   Details of the Maintenance of Dark-fly

In November 1954, the Dark and control stocks of *Drosophila melanogaster*, Oregon-R-S strain, were established, separate from the experimental stock of this strain, which had been cultured with Pearl's synthetic medium for about 6 years

Fig. 1.3 Containers used to keep the fly stocks. Sterilized milk-bottles plugged with silicon plugs were used to rear the flies. The milk-bottles were placed in the larger, light-proof container to rear the Dark-flies

(140 generations) in the Department of Zoology, Kyoto University. This experimental stock had been separated in 1948 from the original stock that had been cultured with kozi (rice malt) medium for a long time in the department. The control and Dark stocks were each divided into three independent lines (a, b and c for the control stock, d, e and f for the Dark stock).

A sterilized milk-bottle plugged with a cotton-ball or silicon plug is used as the container for the flies (Fig. 1.3). The food of the flies is yeast inoculated on Pearl's synthetic medium (Pearl 1926; Osawa et al 1958). It takes about 12 days from the egg to the adult stage in our laboratory conditions, and thus adult flies are transferred every 14–17 days to a new bottle with fresh food.

Bottles of Dark-flies are placed in a light-proof can painted black inside with a blackout curtain cover over the lid (Fig. 1.3), and kept in a temperature-controlled (25 °C) room under the normal light condition. Transfer of the Dark-fly is performed as follows. Firstly, a keep-out notice is hung on the outside of the room door and all the room lights are turned off. Then, the plug of the new empty bottle with fresh medium is pulled out in darkness, and the flies in the old bottle are gathered on the bottom by sharply tapping it against a knee. Immediately after extraction of the plug of the old bottle containing flies, the bottle is tilted upside down, making mouth–mouth contact of the two bottles in order to drop the flies into the new bottle by sharply tapping the bottom of the top bottle. After transfer, the bottles are immediately plugged. During the transfer, the two plugs are pinched between the middle and ring fingers of the respective hand opposite to that holding each bottle, so as not to drop them in the darkness. For confirmation of successful transfer, the presence of flies moving in the new bottle is checked using a very dim red light (640 nm peak wavelength, 0.02 μW/cm²).

In many cases, one line of the *Drosophila* stocks is maintained in two or three parallel sub-lines. As Pearl's synthetic medium does not contain preservatives, the medium in some bottles is occasionally spoiled by the growth of mold. In that case, the flies are discarded, and a new sub-line is established from another sub-line of the same line. Unfortunately, lines a, b, c, d, and e became extinct over time, and only line f of the Dark stock has survived until the present.

1.3   Summary of Experiments Performed So Far

Many experiments have been performed by Mori and his colleagues to study some characters of the Dark-fly. As their interest was focused on the genetic changes that occurred in the Dark-fly line as a result of dark life, the flies to be compared were reared in the same experimental conditions (generally in the light condition) for at least one generation before testing each character. In some cases, statistical analyses performed by them did not suffice independence of data, and thus we rechecked them. The studies of Mori and colleagues on the following phenomena are examined here: (1) response to light, (2) daily emergence rhythm, (3) sensitivity of olfaction, (4) elongation of head bristles, (5) fertility, and (6) developmental rate. Circadian rhythm of locomotor activity, micro-morphology of the compound eye, and genome features of the Dark-fly are treated in Chaps. 2, 3 and 4, respectively.

1.3.1   Response to Light

Response to light is one of the behaviors that are expected to be strongly affected in animals selected by living in the dark. This response has been measured by two criteria: the strengths of phototaxis and photokinesis.

To check the phototactic response of the fruit flies, ten flies were put in a transparent glass tube (3.5 cm in diameter, 40 cm long). After 20 min of dark adaptation, the flies were gathered at one end of the tube by tapping the vertically held tube, and then the tube was held horizontally and suddenly illuminated from the other end by a light parallel to the tube. The tube was immersed in a temperature-controlled water bath set at 25 °C, and the number of flies that passed across the middle line marked on the tube (Fig. 1.4a) within 30 s was counted.

To check the photokinetic response of flies, the same glass tube was used. In this test, flies were not gathered, and the tube was illuminated from the side evenly over its whole length after 20 min of dark adaptation, and the number of flies that moved across any of three lines drawn at equidistant intervals (Fig. 1.4b) during a minute was counted.

The phototactic response measured by the above method includes the photokinetic component, because the effect of the illumination on the level of locomotor

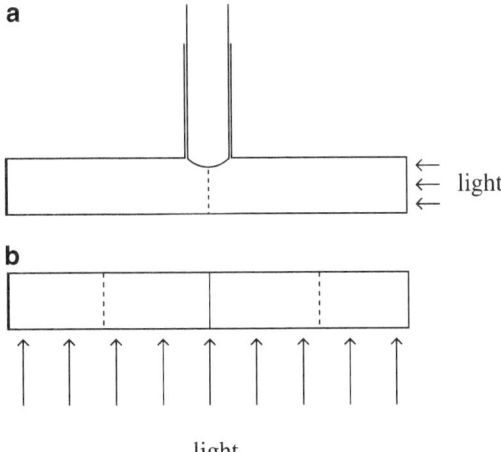

light

Fig. 1.4 Glass tubes used experiments testing response to light. Experiment (**a**): Ten flies were put in a glass tube. In this case, flies were collected in the *left side* of the tube and were illuminated from the *right side* of the tube. The numbers of flies that passed across the midline (*broken line*) within 30 s was counted. Experiment (**b**): Same glass tube as in experiment (**a**) was used. Without collecting the flies, tube was illuminated from the side evenly over its whole length. The number of flies that moved across any of three equidistant lines during 1 s was counted

activity (the photokinetic response) should also change the speed of flies that go toward the light source (the phototactic response). Therefore, the term "phototaxis + photokinesis" was used by Mori and colleagues for this method of testing phototaxis (Mori and Yanagishima 1957), though we adopt simply "phototaxis" for this phenomenon here.

The Dark- and control flies were tested at 39, 51, 80, 82, 108, 135, 168, 202, 304, and 582 generations (Mori and Yanagishima 1959a; Mori et al. 1966; Mori and Imafuku 1982). The Dark-flies to be tested were reared under the same light–dark cycling condition as the control flies for one generation. In the analyses made here, the data from different lines were averaged and compared between the Dark and control flies by Wilcoxon's signed-rank test, after separating them into an "early group" of 39–108 generations, and a "late group" of 135–582 generations, and adjusting the number of generations tested to be five in both groups.

The Dark-flies exhibited a stronger response to light as assessed by both phototaxis and photokinesis (Fig. 1.5). For both behaviors, statistically significant differences were seen between the Dark-fly and control males of both generation groups, and for females of the late generation group. These results indicate that the Dark-flies are clearly sensitive to light, and the correspondence seen between the two light-response behaviors suggests that the photokinetic component is predominant in the light-sensitive behavior of the Dark-flies.

The results obtained in those experiments appear to confirm that some characteristics of organisms are modified according to changes of the environment. If this is

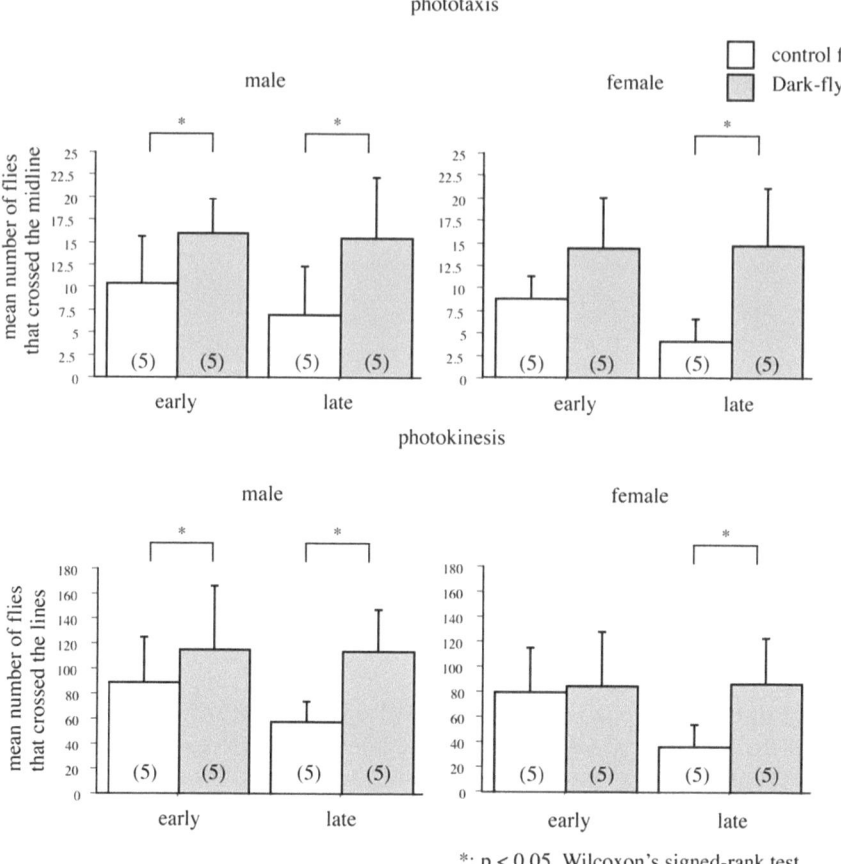

Fig. 1.5 Difference in the light responses of control (male and female) flies and Dark (male and female) flies. *Left panels*: males. *Right panels*: females. *Top panels*: phototaxis. *Bottom panels*: photokinesis. *Numbers in parethesis*: numbers of generations examined. *Error bars* show standard deviation. An *asterisk* indicates a significant difference. The early group of flies were from 39 to 108 generations and the late group of flies ware from 135 to 582 generations

the case, the observed stronger responses to light of the Dark-flies might be diminished if they were returned to normal light conditions for some number of generations. Mori and his colleagues performed experiments to test this possibility by shifting the Dark-flies of 38 generations (the short-term group) to normal light conditions for 1–38 generations (Mori and Yanagishima 1959b), and Dark-flies of 81 generations (the long-term group) to normal light conditions for 11–117 generations (Mori 1983). Then, phototactic responses were compared between the Dark and control flies, testing the same numbers of generations for the two (short-term and long-term) groups: 1, 2, 4, 10, 14, 23 and 38 generations of normal light for the short term group, and 11, 21, 31, 41, 51, 59 and 69 generations of normal light for the long-term group.

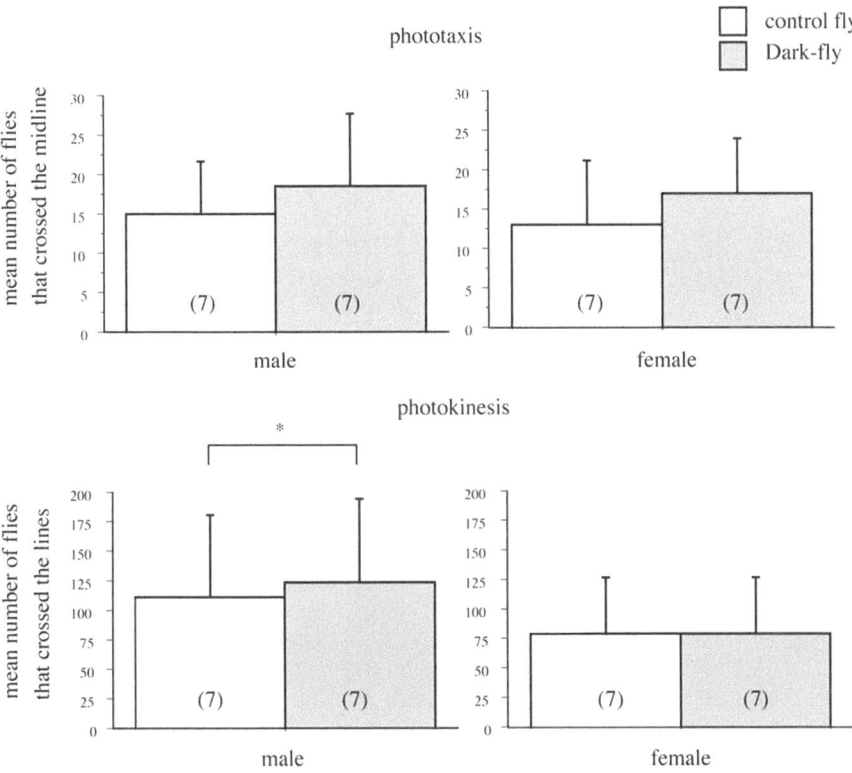

Fig. 1.6 Light responses of short-term group. *Left panels*: males. *Right panels*: females. *Top panels*: phototaxis. *Bottom panels*: photokinesis. *Numbers in parenthesis*: numbers of generations examined. *Error bars* show standard deviation. An *asterisk* indicates a significant difference

   In the short-term group, a significant difference between the Dark- and control flies was detected only for photokinesis of males (Fig. 1.6), whereas in the long-term group, a significant difference was seen between Dark- and control flies for all comparisons tested except for photokinesis of males (Fig. 1.7), the p-value of the latter being marginal (p=0.052). These results suggest that flies kept in darkness for a greater number of generations show greater difficultly in returning to normal light-response behavior being returned to rearing in normal light conditions.

   Finally, Mori (1983) performed crossing experiments between the Dark- and control flies, to check which is genetically dominant: the strong response to light shown by the Dark-fly or the normal response of the control fly. The phototactic response was measured for the four types of crossings. Relatively stronger responses to light tended to be observed in F1 flies, which suggested that the strong response established in darkness was dominant, but the difference was not significant, probably due to the relatively low numbers of individuals examined.

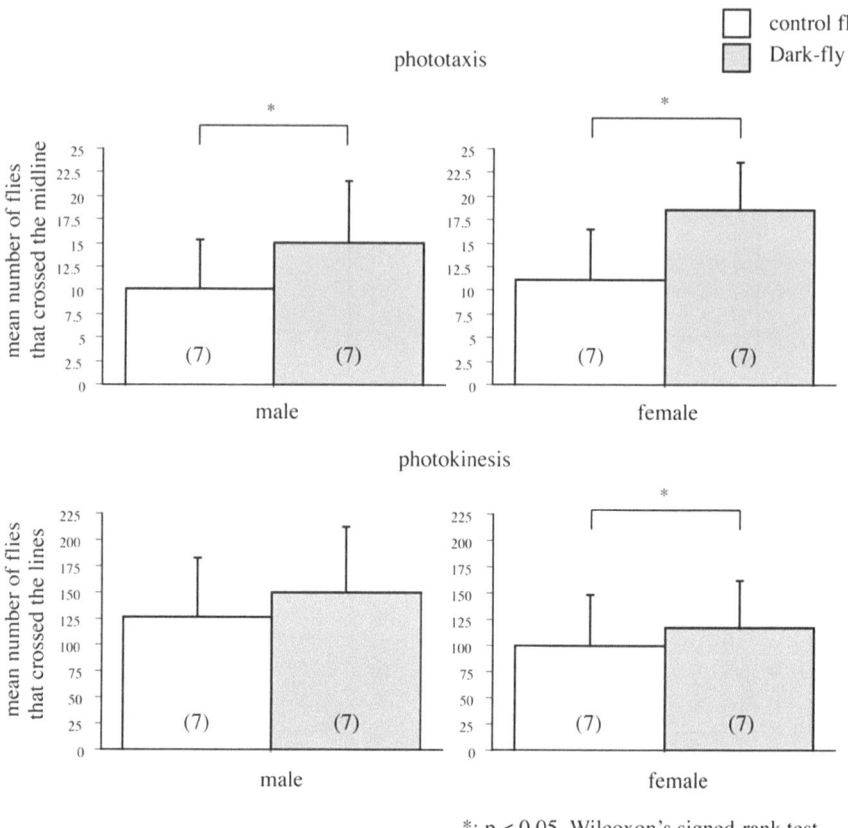

Fig. 1.7 Light responses of long term group. The experiment was similar to that shown in Fig. 1.6, except that flies of the long-term group were used. *Error bars* show standard deviation. An *asterisk* indicates a significant difference

1.3.2   Rhythm of Daily Emergence

Fruit flies emerge mainly from before sunrise until noon, and only in small numbers during the night. The pattern of daily emergence was compared between the Dark- and control flies, by using the fraction of the flies that emerged during the night (18:00–03:00) among the total flies that emerged. To check whether there was a change of the emergence pattern during the course of many generations of life in darkness, the D/L ratio, the Dark-fly's night-emerging fraction divided by the con- trol fly's night-emerging fraction, was calculated at 26, 39, 51, 80, 112, 124, 135, 168, 202, 223, and 238 generations (Mori et al. 1966). The D/L ratio tended to gradually decreases during the course of many generations of dark life, but this tendency was not significant.

Fig. 1.8 Olfactory response
of Dark-fly. The relationship
between convergence ratio
and fermentation period. *Top*
is data for males. *Bottom* is
data for females

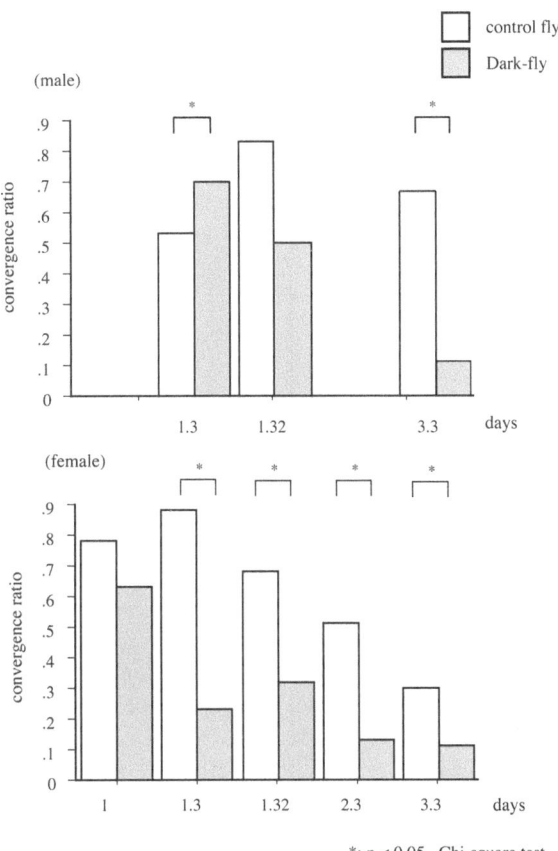

*: p < 0.05, Chi-square test

1.3.3 Olfactory Response

Life in darkness may also induce sensitization of sensory organs other than visual
ones. Olfactory sensory organs are among the candidates of such organs, and their
sensitivity was examined in the Dark-fly as follows. If the Dark-flies have increased
olfactory sensitivity, they might show aversion to heavily fermented medium, as
they are reared on fresh medium. The number of flies that moved from a central
bottle without medium into side bottles with differently fermented media was
counted, and the convergence ratio (the ratio of the number of flies that entered the
bottle with fermented medium to the number that entered any bottle including the
one with non-fermented medium) was calculated. The flies at 230 generations were
used for this experiment (Suzuki 1967).

   The tested flies were always attracted to the fermented medium when it was in an
early stage of fermentation, but this attraction decreased gradually as the duration of
fermentation increased. The extent of the decrease was much greater in the Dark-
flies than in the control flies (Fig. 1.8), indicating that the flies that live in darkness
for over 200 generations become more sensitive to some repellent substances.

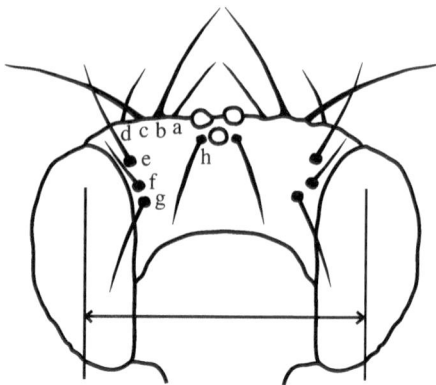

Fig. 1.9 Detailed diagram of the head bristles. The width of the front of the head is represented as the distance between the sides of the *arrow*

1.3.4 Elongation of Head Bristles

Sensory organs related to mechanical stimulations may also change under conditions of darkness. The head bristles of Dark-flies that were reared in normal light conditions for 5 ($D_{583}L_5$) or 30 generations ($D_{583}L_{30}$) after having been reared in darkness for 583 generations were compared with those of the control flies that had been reared in normal light conditions for 588 (L_{588}) or 613 generations (L_{613}) by Imaizumi (1979). Statistically significant elongation was found in most of the head bristles of the Dark-flies examined (Figs. 1.9 and 1.10), namely, *the post verticals* (a in the figure), *anterior verticals* (b), *posterior verticals* (c), *short verticals* (d), *posterior orbitals* (e) and *cellars* (h). Differences were always more notable in females than in males (Imaizumi 1979).

1.3.5 Fecundity

The fecundity of the Dark-flies was compared with that of the control flies (Mori et al. 1966). Ten females of the Dark-flies at 94 generations were made to deposit eggs on Pearl's medium spread on glass slides in normal day light or in the dark environment. The same number of females of the control flies at 92 generations were examined similarly. The numbers of eggs deposited in a week were counted. Both the Dark- and control flies laid significantly more eggs in the dark environment than in the normal day light environment, but flies from the different stocks did not lay significantly different numbers of eggs.

1.3.6 Developmental Rate

The Dark- and control flies at 118 generations were made to deposit eggs. The larvae hatched from the eggs were reared in constant darkness or in normal daylight, and the developmental rates of larvae and pupae were examined (Mori et al. 1966).

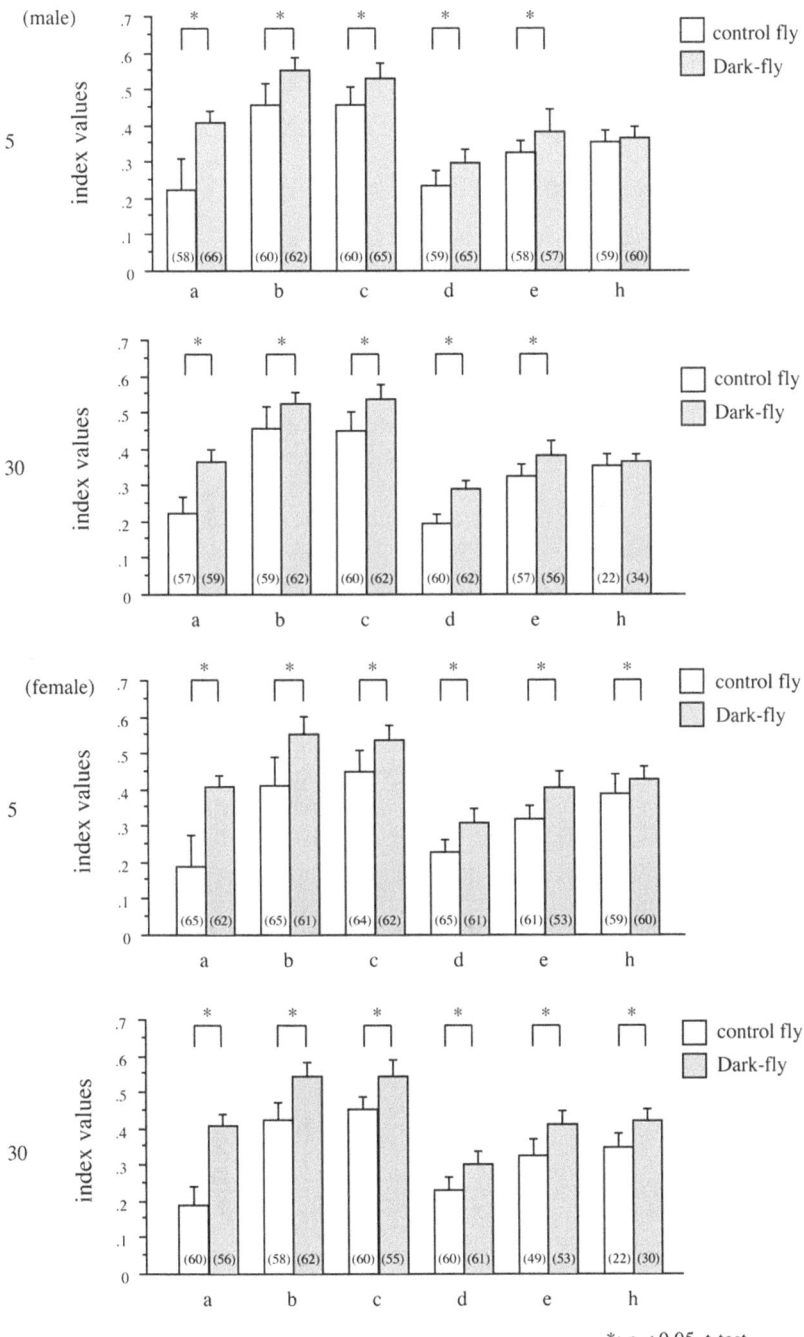

Fig. 1.10 Comparison of lengths of six kinds of bristle. *Top two panels* are data for males. *Bottom two panels* are data for females. The number 5 indicates comparison between $D_{583}L_5$ and L_{588}. The number 30 indicates comparison between $D_{583}L_{30}$ and L_{613}. *Letters in parentheses* are number of bristles examined. *Error bars* show standard deviation. An *asterisk* indicates a significant difference

A significant difference was detected only in larval development between the different environmental conditions to which they were exposed. Thus, developmental rate did not differ between the stocks in either the larval or pupal stage.

1.4   Reflections on the Dark-fly Project

From the results obtained in the series of experiments described above (summarized in Table 1.1), it can be concluded that some characters of the Dark-fly have been modified in the long course of its rearing in total darkness over hundreds of generations. These characters include: sensitization to light (Fig. 1.5), sensitization to olfactory stimuli (Fig. 1.8), and elongation of bristles as tactile receptors (Fig. 1.10). Among them, the last two can easily be imagined to be adaptations to life in darkness, while the first seems to be difficult to understand. Why has the Dark-fly changed in this way?

It is known that animals living in total darkness in nature in environments like the deep sea and the recesses of caves tend to lose their eyes (Park et al 1941; Blume and Günzler 1962; Lamprecht and Weber 1957; Marshall 1957), and become rather insensitive to light (Erckens and Martin 1982). On the other hand, animals living in faint light conditions, such as in an area near the lower limit of the photic zone in the sea, have larger eyes than animals living in normal light conditions (Allee et al. 1949; Marshall 1957), and consequently are expected to be sensitive to faint light (Mori and Yanagishima 1959a). Our Dark-fly has lived in total darkness for about 1,400 generations, but this span is much shorter than that experienced by animals adapted to living in the deep sea or the recesses of caves. It may be thought, therefore, that the nature of our Dark-fly is comparable with that of animals living in faint light conditions. However, our fly has been kept in total darkness, not in faint light conditions. One possible explanation of the characters of our fly is that animals whose optical stimuli are completely removed will, in the initial stage, react to such situations by sensitizing the appropriate organ to detect weak stimuli that might exist, but that if the animals cannot succeed in this for a long time on the evolutionary scale, they will finally discard this reaction and instead acquire diminished sensitivity by degenerating or losing the organ, as seen in cave animals. If this is the case for our Dark-fly, it will gradually become insensitive to light, probably with degeneration of its eyes, in the future.

Table 1.1 Summary of the experiments using "Dark-fly"

Character	Change in Dark-fly
Phototactic response	Sensitive
Photokinetic response	Sensitive
Rhythm of daily emergence	n.s.
Olfactory response	Sensitive
Head bristles	Elongated
Fecundity	n.s.
Development	n.s.

n.s. statistically not significant

Evolution is based on genetic changes. What kind of genetic changes have occurred in Dark-fly? Which of them are responsible for sensitization to light and odor, and for elongation of sensory hairs? Although Dark-fly displays no changes in such phenomena as circadian rhythm or fecundity, cryptic genetic changes might be accumulated before the appearance of phenotypic changes. The major purpose of this project is to investigate what happens at the early stage of evolution. As seen in Chap. 4, approaches using molecular biological techniques would shed light on genetic changes in Dark-fly and would move this project toward a new phase in the future. Our Dark-flies are materials maintained in an "artificial laboratory" to elucidate what occurs in the initial phase of evolution in darkness, a kind of evolutionary "experiment" like those the earth has been conducting in its "natural laboratory" for eons (Paulson and White 1969).

References

Allee W, Emerson AE, Park O, Park T, Schmidt KP (1949) Principles of animal ecology. Philadelphia, W. B Saunders Company

Blume JB, Günzler E (1962) Zur Aktivitätsperidik bei Höhlentieren. Naturwissenscaften 22:525

Erckens W, Martin W (1982) Exogenous and endogenous control of swimming activity in *Astyanax mexicanus* (Characidae, Pisces) by direct light response and by a circadian oscillator. II. Features of time-controlled behaviour of a cave population and their comparison to an epigean ancestral. Z Naturforsch 37c:1266–1273

Imaizumi T (1979) Elongation of head bristles found in a strain of *Drosophila melanogaster*, which have been kept under constant darkness for about 24 years. Jpn J Genet 54:55–67

Lamprecht G, Weber F (1957) Die circadian-Rhythmik von drei unterschiedlich weit an ein Leben unter Höhlenbedingungen adaptierten *Laemostenus*-Arten (Col. Carabidae). Anim Spéléol 30:471–482

Marshall NB (1957) Tiefseebiologie. Gustav Fischer, Jena

Mori (1957a) Variations of *Drosophila* in relation to its environment V: Variations induced in the Pearls medium when transferred from the Kozi medium. Jpn J Genet 32(2):88–99

Mori (1957b) Variations of *Drosophila* in relation to its environment VI. Variations induced in the Kozi medium when transferred from the Pearl's medium. Jpn J Genet 32:277–285

Mori (1983) Variations of Drosophila in relation to its environment VIII. Change of behavior of Drosophila melanogaster seen during 581 generations kept successively in total darkness. Zool Mag 92(2):138–148

Mori S, Imafuku M (1982) Variations of *Drosophila* in relation to its environment VIII. Change of behavior of *Drosophila melanogaster* seen during 581 generations kept successively in total darkness. Zool Mag 91(1):245–254

Mori S, Yanagishima S (1954) Variations of *Drosophila* in relation to its environment III. Variations in the Pearl's medium containing methylene blue. Physiol Ecol 6:43–48

Mori S, Yanagishima S (1957) Variations of *Drosophila* in relation to its environment V: variations induced in the Pearls medium when transferred from the Kozi medium. Jpn J Genet 32(1):57–66

Mori S, Yanagishima S (1959a) Variations of *Drosophila* in relation to its environment VII: Does *Drosophila* change its characters during dark life? Jpn J Genet 34(1):151–161

Mori S, Yanagishima S (1959b) Variations of *Drosophila* in relation to its environment VII: Does dark life change the characters of *Drosophila*? Jpn J Genet 34(1):195–200

Mori S, Yanagishima S, Suzuki N (1966) Influence of dark environment on the various characters of *Drosophila melanogaster* In: Tromp SW, Weihe WH (eds) Proceedings of the 3rd international biometeorological Congress. Biometeorology II, 1–7 September 1963, Pergamon, Oxford, pp 550–563

Osawa W, Matutani K, Tsukude H, Mori S, Yanagishima S, Sato Y, Naka K (1958) Variations of
    Drosophila in relation to the environment II. Variations of *Drosophila melanogaster* in the
    medium containing a sublethal dose of copper sulfate. J Inst Polytech Osaka City Univ Ser D
    9:41–68
Park O, Roberts TW, Harris SJ (1941) Preliminary analysis of activity of the cave crayfish,
    Cambarus pellucidus. J Am Nat 75:154–171
Paulson TW, White WB (1969) The cave environment. Science 165:971–981
Pearl R (1926) A synthetic food medium for the cultivation of *Drosophila*. J Gen Physiol 9:513
Suzuki N (1967) Variation in olfactory response of *Drosophila melanogaster* reared in constant
    darkness. Zool Mag 76:13–20

Chapter 2
Circadian Rhythm of Dark-fly

Abstract Locomotor activity rhythms of Dark-stock flies of *Drosophila melanogaster* kept in complete darkness for 700–1,340 generations were examined. The activity of flies was recorded under the conditions of continuous darkness and of light–dark cycling. The activity rhythm of the experimental dark stock flies observed in continuous darkness was not weakened at 1,340 generations or at 700–900 generations. The degree of diurnality, measured by the ratio of daytime activity to the total activity per day under light–dark cycling, did not differ between the Dark-fly of 700–900 generations and the control fly subjected to normal daylight conditions for the same number of generations. Dark-flies showed a slightly weaker depression of activity in the daytime than control flies did. The possibility of alteration of circadian rhythm in Dark-fly is discussed.

Keywords Activity rhythm • Circadian rhythm • Darkness • Diurnality • *Drosophila melanogaster* • Fruit fly

2.1   Introduction

In 1954, the late Professor S. Mori of Kyoto University launched a long-term experiment in which strains of the fruit fly *Drosophila melanogaster* were maintained over many generations under the condition of complete darkness at a constant temperature to address possible effects of darkness on organisms (Mori and Yanagishima 1959). Since then, several aspects of these flies' behavior, sensory physiology, growth rate, fecundity, and morphology have been examined.

Circadian rhythm is one of characters that are expected to be affected by experimental conditions lacking day–night cycles. Mori et al. (1964) examined the pattern of daily emergence rhythm, measuring the ratio of the number of flies that emerged during nighttime (from 18:00 to 3:00) to the total number that emerged in a day. This ratio was relatively high in Dark-flies, indicating that they had a less distinct emergence rhythm than the control flies from 26 to 135 generations, but the opposite became true thereafter through 238 generations (the last time examined). Circadian rhythm of locomotor activity was compared between Dark and control flies at 1,300 generations, with the result that attenuation of the rhythmicity did not occur in the Dark-fly (Imafuku and Haramura 2011).

Here, we compare some aspects of activity rhythms between Dark and control flies: the strength of rhythmicity in continuous darkness, the degree of diurnality in light–dark conditions, and the activity pattern in the daytime under light–dark conditions.

2.2 Materials and Methods

2.2.1 Flies

Two types of stocks, the Dark and control stocks, were established from Oregon RS strain of *D. melanogaster* in November, 1954 (Mori and Yanagishima 1959). The Dark stock was maintained in complete darkness in a light-proof can (24 cm high, 24.5 cm in diameter) placed in a temperature-controlled room at 25 °C in the Department of Zoology, Kyoto University. The control stock was maintained at the same temperature in an incubator with glass windows through which natural day-light entered, or in a light- and temperature-controlled room with a 12 or 14 h light phase (LD12:12 or LD14:10).

Flies were cultured in ordinary milk bottles (14 cm high, 5.5 cm in diameter) with Pearl's synthetic medium (Pearl 1926; Ohsawa et al. 1958), and transferred to fresh medium to produce the next generation every 2 weeks. All the transfer proce-dures were performed in complete darkness for the Dark stock. The population size of the flies in the bottle fluctuated mostly between 50 and 200.

Flies of the Dark and control stocks from 710 to 918 generations (first group) and from 1,335 to 1,342 generations (second group) were tested. As the control stock unfortunately became extinct at the 1,224th generation on 6 August 2002, a new con-trol stock (K series) was re-established on 11 November 2005 as a derivation of the same strain kept in the *Drosophila* Genetic Resource Center, Kyoto Institute of Technology, where flies had been reared with standard cornmeal medium (https://stockcenter.ucsd.edu/info/food_cornmeal.php) under the conditions of LD12:12 at 23–24 °C. Before coming to the center on 18 December 2002, the stock flies had been maintained under roughly LD11:13 at 22 °C in Bloomington *Drosophila* Stock Center, Indiana University. The newly established light stock was reared with Pearl's medium, and used from 44 to 51 generations in our laboratory, as a control of the second group.

2.2.2 Recording of Activity

A fly placed in a transparent plastic cell (1 × 1 cm, 4.5 cm high) was monitored with a combination system of an infrared light-emitting diode (TLN103A, peak wave-length 940 nm, Toshiba) and a phototransistor (TPS601A, Toshiba) (Fig. 2.1), and

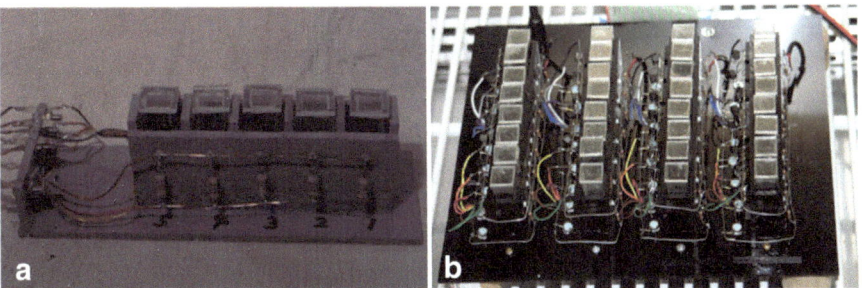

Fig. 2.1 The monitoring part of the activity recorder system. The activity of 5 (**a**) and 32 (**b**) individuals was simultaneously recorded for the first and second group flies, respectively

the on/off information was recorded on a counter IC (MSM5511RS, Oki) for flies of the first group, or directly on a personal computer (OptiPlex GX100, DELL) through an interface (PCI-2703A, Interface Corp.) for flies of the second group.

2.2.3   Data Analysis

The following three characters were compared between Dark and control flies: strength of rhythmicity, ability to entrain to light–dark cycles, and activity pattern in the light phase.

For strength of rhythmicity, a part of data published previously (Imafuku and Haramura 2011) was used. Flies of the first and second groups were examined. Flies of both Dark and control stocks were reared from the egg through the adult stage in LD12:12, and on the 2nd to 4th day of the adult stage they were shifted to continuous darkness and their locomotor activity was recorded. Flies whose record was obtained for at least 7 days were used for further analysis. Firstly, a possible long-term trend was removed by application of the moving average (Tomioka et al. 2003) and then the data were subjected to periodogram analysis (Enright 1965; Sokolove and Bushell 1978). The powers (Liu et al. 1991), as an indicator of the strength of rhythmicity, were compared between the Dark and control flies. In some tests of the second group, flies were grown and maintained with cornmeal medium from the egg to the adult stage, including the whole recording period.

For ability to entrain to light–dark cycles and activity pattern comparison, flies of the first group were used. They were grown in LD12:12 conditions and their activity was recorded in the same condition. As an indicator of entraining ability, the degree of diurnality, the percent ratio of the amount of activity in the light phase to the total amount of activity per day, was calculated. This value is expected to fall between 50 (when the fly is active independently of ambient light conditions) and 100 (when the fly perfectly limits its activity to the light phase). For comparison of activity patterns, peak heights and valley depths were measured.

The data obtained were compared by the Wilcoxon rank sum test using JMP software (version 5.1, SAS Institute).

2.3 Results

2.3.1 Strength of Rhythmicity

Example actograms of activity rhythms of individual flies in continuous darkness are shown in Fig. 2.2. All of them exhibited significant endogenous rhythms with a circadian period of 23.4–24.2 h, though the activity of the control flies of the second

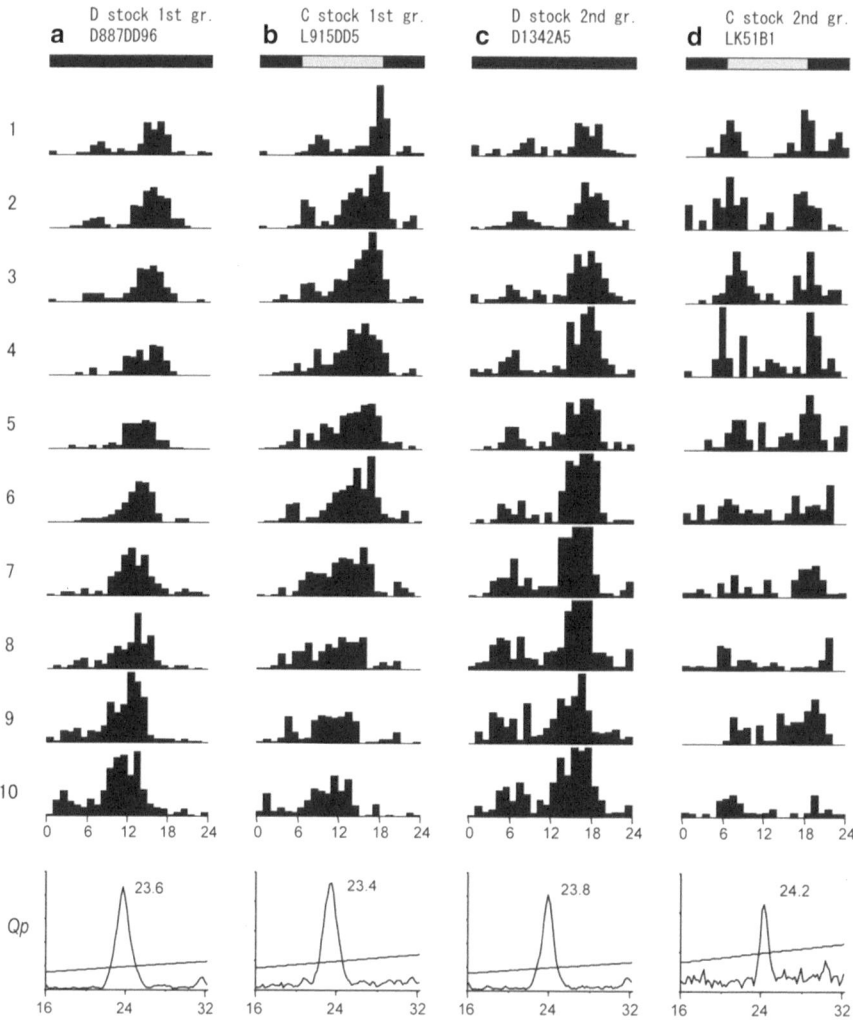

Fig. 2.2 Example actograms of the Dark (**a** and **c**) and control (**b** and **d**) stock flies obtained under continuous dark conditions. Individual codes are shown above the *top bar*, with the generation number in the second to fourth or fifth letters. The generation number for **d** is provided as that for the K series (see Sect. 2.2.1). The *bars* at the *top* indicate the light condition in which the stocks were maintained. The results of the periodogram analysis are shown at the *bottom*, with the most intense period indicated by the number at the *top-right* of the graph, and the significance level indicated by an *oblique line*

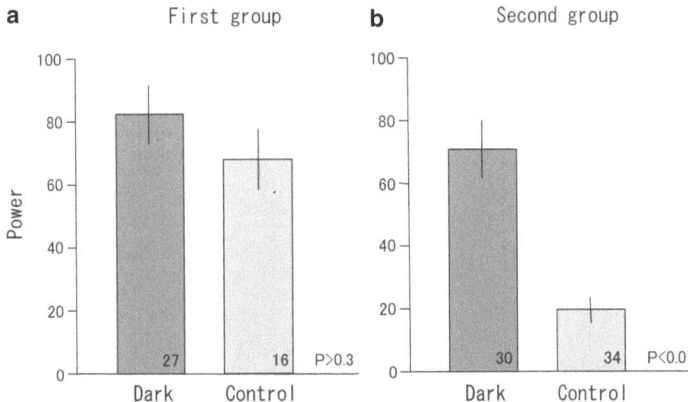

Fig. 2.3 Strength of activity rhythm of Dark and control flies of the first (**a**) and second group (**b**). *Vertical bars* indicate standard error. The number of individuals examined is shown at the *bottom* of the column. The statistical significance of the difference between each pair of values is shown at the *bottom-right* of each panel

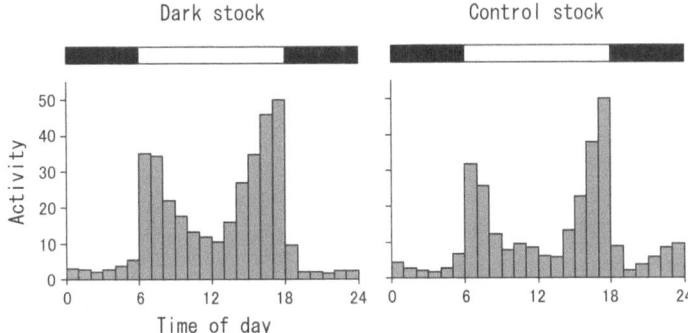

Fig. 2.4 Averaged activity patterns under 12L:12D conditions calculated for 17 Dark- and 15 control flies. Activity is shown as the number of times a fly passed through the infrared beam in the cell per hour

group (Fig. 2.2d) showed a tendency to be somewhat irregular. The strength of rhythmicity was compared between the Dark and control flies, using all the flies examined. It was very similar between the Dark and control flies in the first group (Fig. 2.3a), but clearly differed between them in the second group (Fig. 2.3b).

2.3.2 Entraining Ability

Under the LD12:12 condition, Dark and control flies showed almost the same activity pattern, with bimodal activity in the light phase (Fig. 2.4). The degree of diurnality did not differ between the two stocks (Fig. 2.5a). Thus, Dark flies were entrained to the light–dark cycles as easily as control flies.

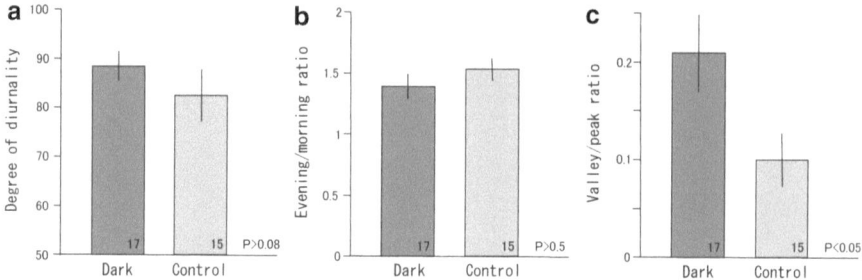

Fig. 2.5 The degree of diurnality (**a**), the ratio of the evening peak height to the morning peak height (**b**), and the ratio of the lowest activity in the light phase to the average of the morning and evening peak heights (**c**). *Vertical bars* indicate standard error. The number of individuals examined is shown at the *bottom* of the column. The statistical significance of the difference between each pair of values is shown at the *bottom-right* of each panel

2.3.3   Activity Pattern

The activity pattern in the light phase was compared between Dark and control flies. To explore whether there was a difference in their activity patterns, the ratio of the evening peak height to the morning peak height was calculated, and no significant difference was found between the two stocks of flies (Fig. 2.5b). To compare the degree of noon time depression, the ratio of the lowest activity in the light phase to the average of the morning and evening peak heights was calculated. These ratios differed significantly between the Dark and control flies ($p < 0.05$, Fig. 2.5c). That is, the depression of the activity of Dark flies was slightly shallower than that of control flies.

2.4   Discussion

In this study, the activity rhythms of *Drosophila* kept in complete darkness for 800 (first group) and 1,300 (second group) generations were compared with that of the control flies exposed to natural daylight conditions during the same number of generations.

Under the LD12:12 condition, Dark flies of the first group exhibited activity patterns similar to those of the control flies: the degree of diurnality, and the ratio of the heights of the two peaks of the bimodal activity pattern did not differ between the flies of the two stocks. In contrast, the pattern of activity during the light phase differed slightly, with a shallower depression around noon in Dark flies. It will be interesting to examine in future generations whether the tendency detected here is further intensified, maintained, or diminished.

In continuous darkness, Dark and control flies of the first group (about 800 generations) exhibited similarly clear rhythmicity, but the Dark and control flies of the second group (about 1,300 generations) showed significantly different degrees of rhythmicity (Fig. 2.3). The latter result, however, is not simply attributable to the number of generations in which Dark flies experienced darkness, because the strength of rhythmicity in the control flies of the second group was clearly lower than that of Dark flies of the same group, and also lower than that of the Dark and control flies of the first group (Fig. 2.3). This lack of clear rhythmicity shown by the control flies of the second group seems to be attributable to the historical conditions of its culturing. Flies of this control stock had been reared with a nutritious medium containing corn and yeast before coming to our laboratory, where they were fed the nutrient-poor Pearl's medium, which is especially lacking in carotenoid (Eguchi 1986). As Dark-flies of the second group reared with this poor medium exhibited a clear activity rhythm, the transition from rich to poor medium that the control flies experienced seems likely to have been a cause of their lack of clear rhythmicity. Independently of what the cause may be, the fact that the Dark-flies subjected to darkness for more than 1,300 generations manifested a clear activity rhythm suggests that a circadian rhythm of *Drosophila* is deep-seated in its body.

Maintenance of circadian rhythm over generations has also been demonstrated by other studies. Rats kept in constant light for 25 generations exhibited a circadian activity rhythm (Browman 1952). *D. melanogaster* kept in light conditions for 600 generations exhibited clear rhythms of eclosion in continuous light or dark conditions (Sheeba et al. 1999).

In contrast to our fruit fly and the animals mentioned just above, some genuine cave animals ("troglobites") are known to be arhythmic or to have weak circadian rhythms. These include the crayfish *Cambarus pellusidus* (Park et al. 1941), the amphipods *Niphargus puteanus* (Günzler 1964), and the fish *Astyanax mexicanus* (Erckens and Martin 1982). The period of darkness these animals have experienced is tremendously long (millions of years). This fact suggests that our Dark-fly may attenuate its rhythmicity or become arhythmic after a significantly long time in the dark condition in the future. A gene controlling the circadian activity rhythm is known in *D. melanogaster* (Konopka and Benzer 1971). Mutation of such a gene may occur at some time in our *Drosophila* population, and the probability of displacement of the normal allele by a mutant allele in the population can be calculated. According to our calculation based on the neutral theory, the number of generations necessary for the occurrence and fixation of a mutant responsible for arhythmia would be approximately 3,000, indicating that within a further 50 years of dark life hereafter, such events might be expected in our Dark-flies (Imafuku and Haramura 2011).

References

Browman LG (1952) Artificial sixteen-hour day activity rhythms in the white rat. Am J Physiol 168:694–697

Eguchi E (1986) Eyes and darkness – evolutionary and adaptational aspects. Zoolog Sci 3:931–943

Enright JT (1965) The search for rhythmicity in biological time-series. J Theor Biol 8:426–468

Erckens W, Martin W (1982) Exogenous and endogenous control of swimming activity in *Astyanax mexicanus* (Characidae, Pisces) by direct light response and by a circadian oscillator. II. Features of time-controlled behaviour of a cave population and their comparison to a epigean ancestral form. Z Naturforsch C 37:1266–1273

Günzler E (1964) Über den verlust der endogenen tagesrhythmik bei dem hölenkrebs *Niphargus puteanus puteanus* (Koch). Biologischer Zentralblatter 83:677–694

Imafuku M, Haramura T (2011) Activity rhythm of *Drosophila* kept in complete darkness for 1300 generations. Zoolog Sci 28:195–198

Konopka RJ, Benzer S (1971) Clock mutants of *Drosophila melanogaster*. Proc Natl Acad Sci USA 68:2112–2116

Liu X, Yu Q, Huang Z, Zwiebel LJ, Hall JC, Rosbash M (1991) The strength and periodicity of *D. melanogaster* circadian rhythms are differentially affected by alterations in period gene expression. Neuron 6:753–766

Mori S, Yanagishima S (1959) Variations of *Drosophila* in relation to its environment VII. Does dark life change the characters of *Drosophila*? Jpn J Genet 34(2):195–200

Mori S, Suzuki N, Yanagishima S (1964) On the origin of pattern of daily rhythmic activity as species characteristics. Physiol Ecol 12:17–27

Ohsawa W, Matutani K, Tsukuda H, Mori S, Yanagishima S, Sato Y, Naka K (1958) Variations of *Drosophila* in relation to the environment. II. Variations of *Drosophila melanogaster* in the medium containing a sublethal dose of copper sulfate. J Ins Polytech Osaka City Univ Ser D 9:41–68

Park O, Roberts TW, Harris SJ (1941) Preliminary analysis of activity of the cave crayfish, *Cambarus pellucidus*. Am Nat 75:154–171

Pearl R (1926) A synthetic food medium for the cultivation of *Drosophila*. J Gen Physiol 9:513–519

Sheeba V, Sharma VK, Chandrashekaran AJ (1999) Persistence of eclosion rhythm in *Drosophila melanogaster* after 600 generations in an aperiodic environment. Naturwissenschaften 86:448–449

Sokolove PG, Bushell WN (1978) The chi square periodogram: its utility for analysis of circadian rhythms. J Theor Biol 72:131–160

Tomioka K, Numata H, Inoue ST (2003) An introduction to chronobiology. Shokabo, Tokyo (in Japanese)

Chapter 3
Compound Eyes of Dark-fly

Abstract The ultrastructure of the compound eye of Dark- and control flies was examined. Flies of both groups reared with nutrient-poor Pearl's medium exhibited a considerable degree of degeneration of the rhabdomeres. When they were reared with nutrient-rich standard cornmeal medium, the rhabdomeres of Dark-fly showed partial degeneration, while those of the control fly showed no degeneration. Thus, nutrients showed a far larger effect on the morphology of the eye than environmental light conditions, which slightly affected on eyes of flies that had experienced many generations of life without light.

Keywords Compound eye • Degeneration • *Drosophila melanogaster* • Fruit fly • Nutrient • Rhabdomere

3.1 Introduction

Visual perception is a process that has been expected to be significantly affected by abnormal light conditions. At the 617th generation of rearing in darkness, the Dark-flies exhibited a "supernormal" structure in the compound eyes: the rhabdomeral microvilli were more densely packed than those of control flies kept in a normal light–dark cycle. The over-developed microvilli were completely destroyed by 3 days' light exposure after emergence (Eguchi and Ookoshi 1981). The same phenomenon was found in the flies at the 796th generation (Eguchi and Arikawa 1985).

The Dark-flies have experienced the dark life for an additional 500 generations since the last investigation was performed. In this period, the stock of original control flies (Control 1) unfortunately became extinct. We therefore introduced a new pseudo-control stock of the same strain, obtained from the *Drosophila* Genetic Resource Center, Kyoto Institute of Technology, where it had been maintained under the normal light–dark condition and fed standard cornmeal medium (Control 2). In our laboratory, the new control stock was fed our medium, Pearl's synthetic medium (for the composition, see Methods). In the present study, we explored changes in the rhabdomere structure of flies kept in the dark for more than 1,300 generations, along with possible effects of the diet that might have led to eye degeneration in previous generations (Eguchi 1986).

3.2   Materials and Methods

3.2.1   Animals

Flies of three different groups were examined: Dark-flies at the 1,308th generation (DD flies), control flies (LL flies), and offspring from DD flies that were shifted to normal light conditions for five generations prior to the present observations (DL flies).

All flies were fed with Pearl's medium (Pearl 1926). The DD flies were kept in complete darkness except when we changed the medium once every 2 weeks (during this procedure the flies were briefly illuminated with dim red light (about 700 nm) to verify their successful transfer to fresh medium). We also used the standard cornmeal medium (https://stockcenter.ucsd.edu/info/food_cornmeal.php) to check the effect of the diet.

3.2.2   Electron Microscopy

For transmission electron microscopy (TEM), heads of the flies were hemisected in a fixative [2.5 % glutaraldehyde, 2 % paraformaldehyde in 0.1 M sodium cacodylate buffer (pH 7.3, CB)] on ice and prefixed in the fixative overnight at 4°C. After washing with 0.1 M CB, the tissues were postfixed in 2 % osmium tetroxide in 0.1 M CB for 2 h at room temperature, washed with distilled water, and dehydrated in a graded series of acetone. The tissues were then infiltrated with propylene oxide and embedded in Quetol 812 resin. Ultrathin sections cut with a diamond knife were double stained with uranyl acetate and lead citrate, and observed with an electron microscope (Hitachi H7650).

3.3   Results and Discussion

TEM observations revealed that all of the DD, LL and DL flies kept on Pearl's medium showed a considerable degree of degeneration in the photoreceptors: microvilli of rhabdomeres were largely disordered and frequently protruded into the cell body, swelling to form large vacuoles (Fig. 3.1, middle panels). This result is not consistent with the previous results reported by Eguchi and Ookoshi (1981), in which the DD flies at the 617th generation expressed "super-normal" structures

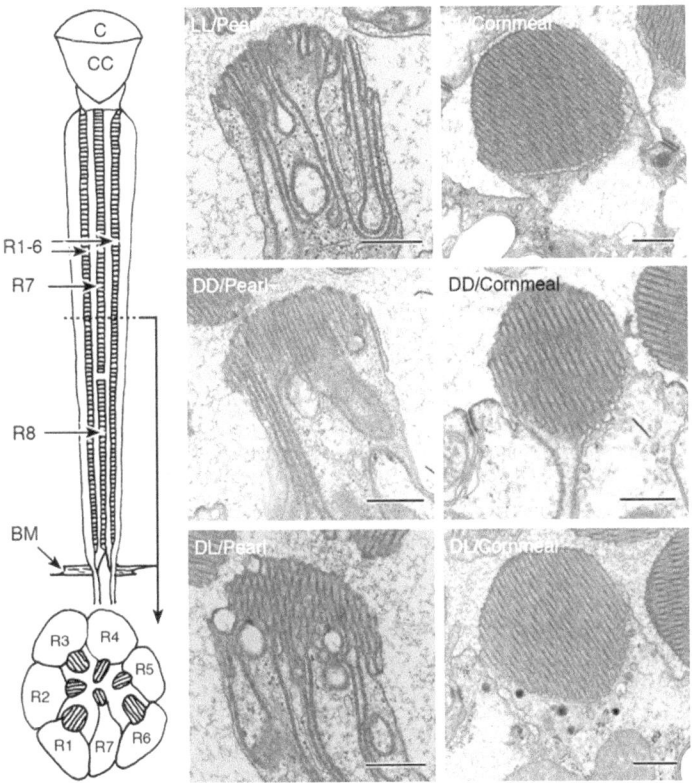

Fig. 3.1 Diagram of an ommatidium in longitudinal (*left top*) and transverse (*left bottom*) views. *C* cornea, *CC* crystalline cone, *R1-8* photoreceptors, *BM* basement membrane. Six TEM pictures show single rhabdomeres from transverse section of an ommatidium of LL (*top*), DD (*middle*) and DL (*bottom*) flies. Flies were cultured with Pearl's medium for 617 generations (*middle column*) or with Pearl's medium (617 generations) followed by cornmeal medium for two generations (*right column*). Scale bars, 0.5 μm

with larger rhabdomeres with more densely packed microvilli that were thinner than those of LL control flies.

Following the first observation in the present study, all three groups of flies, DD, LL and DL, were reared with the standard cornmeal medium instead of Pearl's medium for two generations. The standard cornmeal medium-reared LL flies showed normal structure of the rhabdomeres, and the DD flies and DL flies exhibited only a minor degree of degeneration (Fig. 3.1, right panels).

Figure 3.2 summarizes our present results together with those published previously on the dark-reared flies, based on which we draw the following two conclusions. Firstly, flies kept in darkness for several hundred generations expressed, in the early stage, morphological changes consistent with sensitization to light, and therefore as called "supernormal", and then shifted to having degenerated eyes. This temporal pattern may be comparable with that observed in deep sea fishes, which

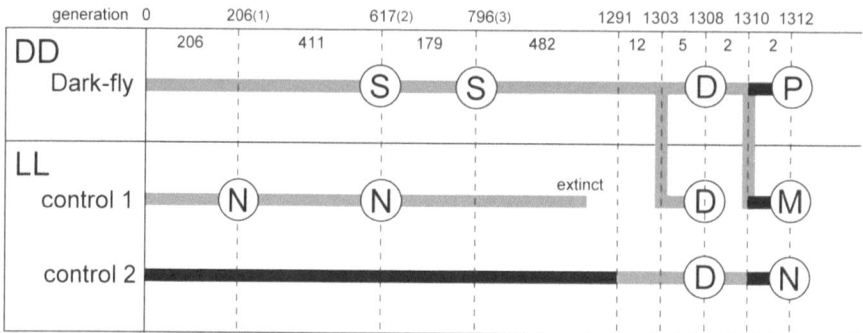

(1) Mori et al. 1966, (2) Eguchi & Ookoshi 1981, (3) Eguchi & Arikawa 1985.

Fig. 3.2 Changes in ultrastructure of the compound eyes of *Drosophila* observed at different generations. DD and LL in the *left column* indicate continuous darkness, and normal light and dark conditions, respectively. Control 1 means the original control stock, and control 2 a newly established control stock. *Gray and black bars* indicate the Pearl's and cornmeal media, respectively, with which flies were cultured. Letters in *circles* are: *N* normal, *S* supernormal, *D* degenerated, *P* partially damaged, and *M* minor degeneration. Numerals below the *top line* indicate generation numbers between the successive *vertical broken lines*

appear to have gradually adapted to deeper and darker environments on the scale of the evolutionary time: fishes living in water around depths of 1,000 m developed highly sensitive eyes, while those inhabiting water deeper than this level underwent degeneration of their eyes (Eguchi 1986).

Secondly, as is well known, diet has significant effects on the rhabdomere structure. Degeneration of rhabdomeres in DD flies can be almost completely prevented by culturing the flies with the standard cornmeal medium for two generations. The newly introduced control (LL) flies that were reared with cornmeal medium (Control 2) showed serious degeneration of the rhabdomeres within 17 generations of culturing with Pearl's medium, but the degeneration was completely recovered by culturing with cornmeal medium for two generations. It has been demonstrated that the quantity of the visual pigment chromophore 11-*cis* 3-hydroxyretinal in flies reared with Pearl's medium was as low as 5–10 % of that in flies fed cornmeal medium (Eguchi 1986). This is in fact the result of the low vitamin A content of Pearl's medium. The observed degeneration in the present study was therefore most likely due to vitamin A deprivation (Goldsmith et al. 1964; Carlson et al. 1967; Brammer and White 1969; Suzuki et al. 1993).

Here one question arises: why did the original LL flies (Control 1) that had been reared with the vitamin A-deprived medium not show rhabdomere degeneration at the 206th and 617th generations? Long-lasting culture with vitamin A-deprived medium may have selected individuals whose eyes do not degenerate in such severe dietary conditions.

Finally, our Dark-flies seem to be in the process of undergoing adaptation to a dark environment. In addition to morphological modifications, physiological modifications related to dark life, such as sensitivity to light, vibration, and tactile and chemical stimuli will be intriguing problems to tackle in the future.

References

Brammer DJ, White RH (1969) Vitamin A deficiency: effect on mosquito eye ultrastructure. Science 163:821–823

Carlson SD, Steeves HR III, VandeBerg JS (1967) Vitamin A deficiency: effect on retinal structure of the moth *Manduca sexta*. Science 158:268–270

Eguchi E (1986) Eyes and darkness – evolutionary and adaptational aspects. Zoolog Sci 3:931–943

Eguchi E, Arikawa K (1985) Structure and function of the compound eyes of *Drosophila melanogaster* cultured in darkness for about 800 generations. Zoolog Sci 2:867

Eguchi E, Ookoshi C (1981) Fine structural changes in the visual cells of *Drosophila* cultured in darkness for about six hundred generations. Annot Zool Jpn 54:113–124

Goldsmith TH, Barker RJ, Cohen CF (1964) Sensitivity of visual receptors of carotenoid-depleted flies: a vitamin A deficiency in an invertebrate. Science 146:65–67

Mori S, Yanagishima S, Suzuki N (1966) Influence of dark environment on the various characters of *Drosophila melanogaster*. In: Tromp SW, Weihe WH (eds) Proceedings of the 3rd international biometeorological congress. Biometeorology II. Pergamon, Oxford, pp 550–563

Pearl R (1926) A synthetic food medium for the cultivation of *Drosophila*. J Gen Physiol 9:513–519

Suzuki E, Katayama E, Hirosawa K (1993) Structure of photoreceptive membranes of *Drosophila* compound eyes as studied by quick-freezing electron microscopy. J Electron Microsc (Tokyo) 42:178–184

Chapter 4
Genome Features of Dark-fly

Abstract Recent progress in genome science enables us to determine the whole genome sequences of laboratory-evolved organisms and to address the molecular mechanisms underlying environmental adaptation. We determined the whole genome sequence of Dark-fly and identified many genomic alterations in its genome. Analysis of the population genome structure revealed that about 5 % of genome regions were possibly selected during the Dark-fly history, and that 241 genes in those regions carry mutations. In addition, we identified 28 nonsense mutations in the Dark-fly genome, which probably disrupt or severely affect the function of the gene product. These results revealed unique features of the Dark-fly genome and provided a list of potential candidate genes involved in environmental adaptation. These candidate genes include a light receptor, olfactory receptors, and enzymes related to detoxification and neural development.

Keywords *Drosophila melanogaster* • Environmental adaptation • Genome • InDel (insertion and deletion) • NGS (next generation sequencing) technology • SNP (single nucleotide polymorphism)

4.1 Introduction

4.1.1 Dark-fly as a Model Organism for Studying Environmental Adaptation

Organisms often display beautiful traits adaptive for their environments. How organisms come to possess such adaptive traits is a fundamental question for evolutionary biology. As one answer, Charles Darwin (1809–1882) proposed the natural selection theory: adaptive traits are selected and consequently become prevalent in a population during its history (Darwin 1859). This concept is widely accepted, but

N. Fuse et al., *Evolution in the Dark: Adaptation of* Drosophila *in the Laboratory*,
SpringerBriefs in Biology, DOI 10.1007/978-4-431-54147-9_4, © The Author(s) 2014

its molecular mechanism is not fully understood. To address this issue, experimental evolution studies utilize model organisms evolved in defined environments in the laboratory. Previous experimental evolution studies observed genomic alterations under environmental selection and evaluated the effects of genes on fitness (Barrick et al. 2009; Kishimoto et al. 2010; Chou et al. 2011; Khan et al. 2011). However, those studies generally utilized unicellular organisms, such as bacteria or yeast, because of their short generation times and relatively small genomes. Recent progress in genome science, as represented by next-generation sequencing (NGS) technology, has changed the situation by enabling us to determine the whole genome sequences of sexual organisms from enormous output data (Metzker 2010). NGS is now starting to be used to characterize the whole genome sequences of laboratory-evolved organisms.

In 1954, Syuiti Mori (1912–2007) started an experiment of maintaining a fruit fly, *Drosophila melanogaster* strain (Oregon-R-S strain) in constant dark conditions (Mori 1986; Ashburner et al. 2005). As of 2012, this fly line, designated Dark-fly, has been reared in darkness for 57 years (1,400 generations). Although it has been shown that Dark-fly displays some interesting characters, such as strong phototaxic behavior and lengthened head bristles (see Chap. 1), it is still unclear if Dark-fly is really adapted for living in the dark. Unfortunately, the control sister lines were lost during the rearing history, and only one of three replica lines reared in the dark (fD line) has survived until now (Fig. 4.1). Therefore, it is impossible to compare Dark-fly directly with the control sisters. Nevertheless, Dark-fly is a unique organism reared long-term in a dark environment, and accordingly can be utilized for analyzing traits and genes involved in environmental adaptation. Furthermore, Dark-fly has been reared with a minimal medium, called Pearl's medium (Mori and Yanagishima 1959). There is a considerable possibility that poor nutrient conditions influence the selective pressure in dark environments. Thus, Dark-fly might be useful for analyzing interactive effects of environmental factors on selection, which probably occur in nature.

We measured the fecundity and longevity of Dark-fly to examine its adaptation in the dark, and performed whole-genome sequencing for Dark-fly using NGS technology (Izutsu et al. 2012). Before describing our study, we will briefly introduce previous molecular studies regarding dark-dwelling organisms and laboratory-evolved *Drosophila* lines to clarify the position of our Dark-fly project.

4.1.2   Organisms Living in Darkness

If you dig underground or go inside a cave, you will find many organisms living there. These dark-dwelling organisms might show some odd morphologies noticeable at a glance, such as degenerated eyes and albino skin. You might therefore imagine that these organisms have evolved in the dark by discarding useless tissues or organs. What happens at the molecular level during their evolution?

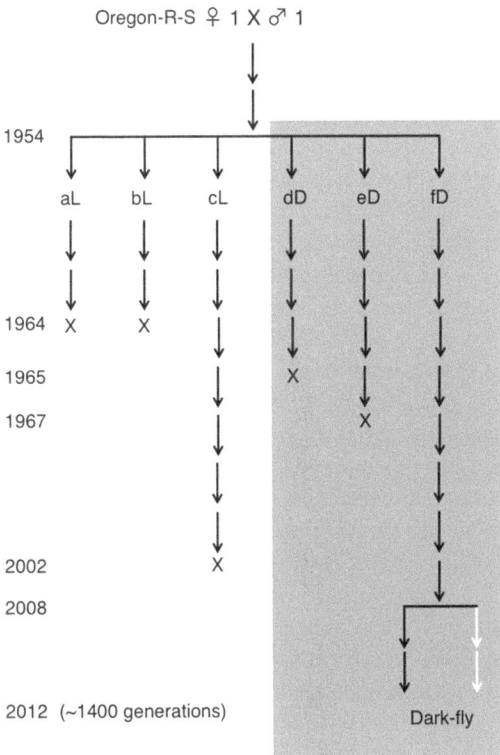

Oregon-R-S ♀ 1 X ♂ 1

Fig. 4.1 History of Dark-fly. In 1954, a fly population derived from one pair of Oregon-R-S flies was divided into six populations. Three of them (aL, bL and cL populations) were reared in normal light–dark cycling conditions and the remaining three populations (dD, eD, and fD populations) were reared in constant dark conditions. Unfortunately, all of the L lines were lost by 2002. The dD and eD lines were lost in 1965 and 1967, and only the fD line has been maintained until now. In 2008, we started to rear the fD line and designated it "Dark-fly." We have maintained Dark-fly in a minimum medium as done before (*black lines*), and in a standard cornmeal medium (*white lines*) in parallel. The population size of Dark-fly has not been controlled but has usually been about 100 flies each in several culture vials

One notable dark-dwelling organism is the African naked mole rat, *Heterocephalus glaber*, which lives in tunnels underground. These rodents display some intriguing features in addition to degenerated eyes and naked skin (Edrey et al. 2011). While their habitat conditions such as low oxygen and complete darkness seem severe, they exhibit extremely long longevity (about 30 years), which is ten times longer than that of mice. They are also resistant to oxidative stress and cancer formation. Tens or hundreds of naked mole rats constitute a eusocial colony with a breeding queen, similar to the colonies of termites and bees. These features have attracted much interest in biological, ecological and medical research fields. Recently, the whole genome sequence of the naked mole rat was determined and has highlighted some genes potentially involved in the responses to low oxygen and darkness (Kim et al. 2011).

The cavefish *Astyanax mexicanus* lives in various Mexican caves, and has evolved independently in different caves from closely related ancestors living in surface waters (Jeffery 2001). Since cavefish in different caves share some morphological characters (degenerated eyes and albino skin), this species provides a model system for investigating convergent evolution under common environmental conditions. Molecular mechanisms for the traits of cavefish have been extensively studied. Genes involved in the albino skin were successfully identified using quantitative trait loci (QTL) analyses, and were found to include genes related to the pigmentation pathway (Gross et al. 2008, 2009). On the other hand, genome alterations responsible for the eyeless trait have not been determined yet. So far, it has been shown that the Shh gene, which encodes a signaling molecule that defines the midline of the craniofacial structure, is expressed more widely in cavefish than in the surface-dwelling fish (Yamamoto et al. 2004). It was suggested that the wider expression of Shh causes a widened jaw at the expense of eyes, and might represent the combination of simultaneously gaining a useful morph and losing a useless morph (Yamamoto et al. 2009). Since genomic loci linked to the eyeless phenotype do not include the Shh gene (Protas et al. 2007), it is expected that unknown gene(s) regulating the expression of Shh might have evolved in cavefish. Another intriguing issue is whether cavefish exhibit unique behaviors. Indeed, cavefish respond to water vibration and approach the vibration-source, whereas surface fish do not (Yoshizawa et al. 2010). As a consequence, cavefish can catch prey (brown shrimp) in the dark more successfully than surface fish, and thus these behavioral traits would be advantageous for living in the dark. Furthermore, cavefish display prolonged longevity (Jeffery 2008), suggesting that their longevity might be due to low metabolism under poor nutrient conditions in the cave.

Historically, many researchers have been fascinated by the odd morphology of dark-dwelling organisms. One of them was Fernandus Payne (1881–1977). He studied the morphology and behavior of cave-dwelling organisms (fish and lizard). In 1907, Payne joined Thomas Hunt Morgan's laboratory at Colombia University, and started to rear *Drosophila* lines in a completely dark environment. Morgan suggested that Payne undertakes this project in order to test Lamarck's theory, known as "inheritance of acquired characters." Payne observed that flies reared in the dark for 69 generations did not show any changes of eye morphology or function (Payne 1911; Ashburner et al. 2005), and this observation is taken as one piece of counter-evidence against Lamarck's theory. Later, Morgan noticed the advantages of flies for studying heredity, and accordingly established the basis of *Drosophila* genetics (Ashburner et al. 2005). Thus, Payne's original "Dark-fly" project contributed to establishing current status of *Drosophila* as an excellent model organism, for which there are abundant sources of information and resources. We made full use of the power of this model organism to characterize our novel "Dark-fly," and analyzed its genome using NGS technology.

4.1.3   Drosophila *Research Using NGS Technology*

NGS technology, using sequencers such as the Illumina Genome Analyzer or Roche 454 system, outputs only a short sequence for each read (30–300 nucleotides), but produces tens or hundreds of millions of the reads in one experiment (Metzker 2010). Such enormous output data enables us to determine the whole genome sequences for organisms, even for human (about 3 Gb genome). NGS technology has been applied recently in some studies of *Drosophila*. Burke et al. selected heterogenous *Drosophila* populations for accelerated development in the laboratory (Burke et al. 2010). After 600 generations, the selected lines develop from egg to imago significantly faster than the unselected lines. Using NGS, they performed whole-genome sequencing for these lines and thereby identified many single nucleotide polymorphisms (i.e., nucleotide sequences altered relative to the reference sequence: SNPs). Some SNPs are homogenous, and thus are fixed in the population, while other SNPs retain heterogeneity in the population genome. Further analyses revealed that homogenous SNPs are accumulated at several regions on chromosomes and some of those regions are shared between the independently selected lines. This phenomenon suggests the occurrence of a "genetic sweep" in evolution (i.e., extension of homogenous SNPs in the population genome through a sweep of selected allelic sequences) (Hermisson and Pennings 2005). A fraction of the prevalent homogenous SNPs are likely to be involved in the selected traits (in this case, related to accelerated development). The selected genome regions constitute 450 genes, some of which are related to larval development, imaginal disc development, and metamorphosis. Although functional analyses of these SNPs remain as future issues, the findings of Burke et al. have shown some of the genome dynamics that occur in laboratory evolution.

In another study of evolution in the laboratory, Zhou et al. reared *Drosophila* populations in hypoxic conditions (Zhou et al. 2011). In those experiments, normal flies all died in the 4 % oxygen condition, but some survived in 8 % oxygen (normal air contains 21 % oxygen). They initially maintained mixed populations of *Drosophila* lines in the 8 % oxygen condition and continued to maintain them in gradually decreased oxygen conditions. After 180 generations, the selected fly lines could survive even in the 4 % oxygen condition. Then, the hypoxia-tolerant lines together with normal lines were subjected to whole-genome sequence analysis using NGS. Zhou et al. (2011) computed allelic differences between populations and successfully identified the SNPs selected in hypoxia-tolerant populations. Interestingly, some of the SNPs were mapped to 12 genes related to the Notch signaling pathway, which functions in various aspects of development. It was previously shown that the hypoxia-response pathway in vertebrate cultured cells is integrated with Notch signaling (Gustafsson et al. 2005). Zhou et al. (2011) confirmed that Notch signaling similarly contributes to the hypoxia-responses of fly.

It is generally considered that complex traits of organisms, such as morphology, behavior and metabolism, are controlled coordinately by multiple genes. To understand the gene networks that operate in complex traits, Mackay and colleagues established various inbred *Drosophila* lines from a natural population (the DGRP lines) (Mackay et al. 2012). They then determined the whole genome sequence of each line using NGS and quantitatively measured many traits, such as aggressiveness and chill-coma recovery. Their statistical analyses comparing genomes and traits in each line demonstrated causal links between multiple genes and complex traits.

Thus, NGS technology has been applied for *Drosophila* research to analyze genome dynamics during population history, signaling pathways in the response to the environment, and gene networks in complex traits. As a result, we are clearly progressing toward understanding the molecular nature of environmental adaptation, though many questions remain unanswered. Do various genes have small effects, or a few genes have large impacts, to achieve environmental adaptation? What kinds of mutations (i.e., alteration of protein function or gene expression) contribute to environmental adaptation? Losing a useless trait or gaining a useful trait: which one happens first for environmental adaptation? To address the molecular nature of environmental adaptation, we utilize Dark-fly. Dark-fly is a model organism reared long-term in a dark environment, and accordingly can be utilized for analyzing traits and genes involved in dark-adaptation. We will now introduce our study on Dark-fly (Izutsu et al. 2012).

4.2   Traits of Dark-fly

4.2.1   Fecundity of Dark-fly

We first asked whether Dark-fly shows successful reproduction in dark conditions, since reproductive success is one of the measures of adaptation. Although Mori and others previously observed normal fecundity in Dark-fly (see Sect. 1.3.5), we again measured fecundity of Dark-fly. Since 2008, we have maintained Dark-fly in a minimum medium (Pearl's medium) as done before, and in a standard cornmeal medium in parallel. To observe the effect of dark environment more directly, we used flies reared in the cornmeal medium for more than ten generations. We also used the Oregon-R-S strain, which was obtained from a stock center, as a control wild-type line, because Dark-fly originated from that strain (Mori and Yanagishima 1959). For the experiment, adult flies were placed in a light–dark cycling (12-h: 12-h; LD), constant light (LL) or constant dark (DD) condition for 3 days and the offspring grown in the cornmeal medium were counted. Oregon-R-S produced approximately 38 offspring/female during 3 days irrespective of whether the flies were tested in the LL, LD, or DD condition (Fig. 4.2a). In contrast, Dark-fly produced significantly more offspring in the DD condition than in the LL condition (mean 42.6 in DD versus 38.6 in LL). A tendency toward relatively high fecundity in the DD condition

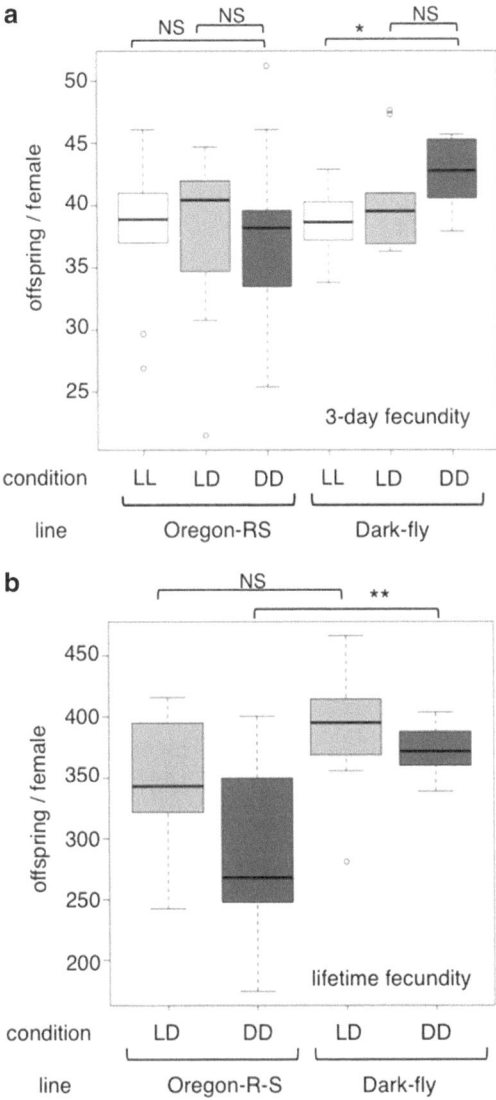

Fig. 4.2 Fecundity of Dark-fly and Oregon-R-S. (**a**) Three-day fecundity (offspring/female) of Dark-fly and Oregon-R-S in LL, LD and DD conditions is shown by *box plots*. *Boxes* and *median lines* represent inter-quartile range and median values of data, and *vertical lines* represent minimum and maximum values of data within 1.5-fold of the inter-quartile range. *Circles* indicate values of outliers. *Asterisk* indicates statistically significant difference (FDR-adjusted p-value <0.05, Welch t-test, n=10, total 100 females). NS means not significant. (**b**) Lifetime fecundity (offspring/ female) of Dark-fly and Oregon-R-S in LD and DD conditions is shown by box plots in a similar manner to that in (**a**). *Double asterisks* indicates statistically significant difference (p-value <0.01, Welch t-test, n=10, total 100 females)

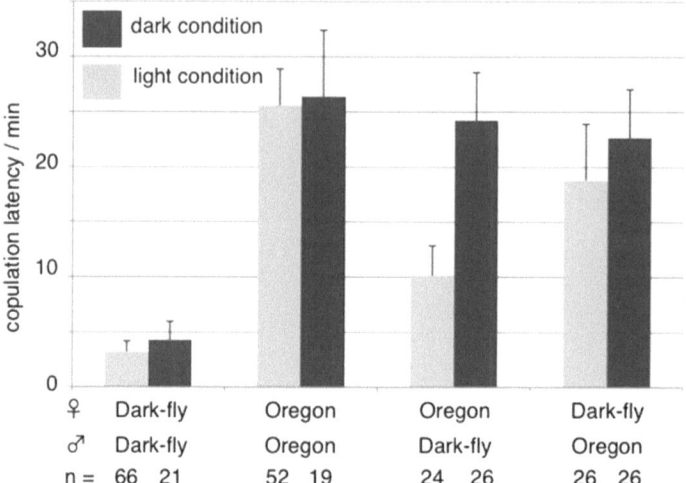

Fig. 4.3 Sexual behavior of Dark-fly. Time latency from starting a courtship behavior until achieving copulation was measured for fly pairs as indicated below the graph. Mean times (minutes) are represented as *bars with lines* (standard errors). The experiments were performed in light (*grey*) and dark (*black*) conditions. Numbers of observations (n) are indicated below the graph. Data showing more than 1 h until copulation were omitted for mean calculation. Such cases constituted less than 20 % of total observations

was also observed when compared with the LD condition (mean 40.3 in LD). These results suggest that Dark-fly produces many offspring in dark conditions over a period of 3 days, but Oregon-R-S does not show such an advantage in the dark.

We next examined the fecundity over a fly's lifetime. Dark-fly produced a similar number of offspring over its lifetime in LD and DD conditions (Fig. 4.2b). This suggests that the reproductive ability of Dark-fly per se is not altered in the dark, but rather Dark-fly produces more offspring early during the mating period (during the first 3 days) in the dark. Oregon-R-S as well as Dark-fly produced approximately 300 offspring/female over its lifetime. It seems that Oregon-R-S decreased the number of offspring produced in the DD compared to the LD condition, but Dark-fly maintained it. Consequently, Dark-fly produced significantly more offspring than Oregon-R-S in the DD condition (mean 373 for Dark-fly versus 293 for the Oregon-R-S). The decreased fecundity of Oregon-R-S in the dark appears to be partly due to decreased adult viability, as described the next section (see Sect. 4.2.2).

The early reproduction of Dark-fly would be advantageous in the laboratory routine of fly maintenance. The early reproduction could be achieved via various traits of the fly, for example, egg-laying ability or mating behavior. To examine mating behaviors of Dark-fly, we observed the behaviors of virgin females and males put together as pairs into an arena (diameter 3.5 cm, depth 1 cm, volume 10 cm^3) and measured the latency time from starting courtship behavior (wing vibration of the male) until achieving copulation. Dark-fly males and females copulated more quickly than the Oregon-R-S pairs (Fig. 4.3: mean 3.2 min for Dark-fly versus 25.6 min for Oregon-R-S), suggesting that mating behaviors might be stimulated in the Dark-fly pairs: males might be more active for courtship and/or females might more easily accept males. Since the quick copulation of Dark-fly was observed in light conditions as well as in dark conditions (Fig. 4.3), the quick copulation alone would not account for the early reproduction in the dark. However, we speculate that stimulated sexual behavior contributes to the early reproduction via re-courtship after failure and/or via repeated mating. We also examined behaviors of heterologous pairs, such as a Dark-fly female and an Oregon-R-S male, and vice versa. The heterologous pairs, especially the pairs of an Oregon-R-S female and a Dark-fly male, achieved relatively quick copulation in light conditions, but this acceleration was completely abolished in dark conditions (Fig. 4.3). Thus, in the dark, only the pairs of a Dark-fly male and female achieved quick copulation. These results suggest that some specific interaction(s) between the Dark-fly pair stimulate sexual behavior, and that the interaction works even in darkness. Mating behavior is controlled by multiple sensory inputs, such as sight, smell, taste and sound (Billeter et al. 2006; Greenspan and Ferveur 2000). One hypothesis is that Dark-fly might be more sensitive to sensory signals, for example, sexual pheromone(s) and the courtship song.

4.2.2   Longevity of Dark-fly

Next, we compared longevity between Dark-fly and Oregon-R-S. When males and females were reared together, Oregon-R-S and Dark-fly males showed similar viability (Fig. 4.4a) but Dark-fly females survived longer than Oregon-R-S females in either the LD or DD condition (Fig. 4.4b). Females of both lines survived longer in the LD condition compared to the DD condition. However, remarkably, Oregon-R-S females gradually died in the DD condition (Fig. 4.4b, solid light blue line), but Dark-fly females did not show such gradual death (Fig. 4.4b, solid dark brown line). Consequently, the 50 % survival period in the DD condition was 43 days for Dark-fly and 24 days for Oregon-R-S. It is unlikely that Dark-fly possesses extraordinary longevity, because Dark-fly virgin females showed shorter longevity than Oregon-R-S virgin females (Fig. 4.4c). Even more surprisingly, Dark-fly virgin females showed shorter longevity than the mated ones (Fig. 4.4b, c, dark brown lines).

Our results suggest that Oregon-R-S females gradually died in dark conditions, while Dark-fly females did not show such gradual death. This phenomenon is probably a complex consequence not easily explained, but it might be related to the fact that Dark-fly females retain longevity after mating. Reproduction is generally a cost for longevity (Weinert and Timiras 2003), and in accord with this, Oregon-R-S virgin females showed much longer longevity than the mated ones. Dark-fly females might not have such a cost of reproduction. The cost of mating for females is thought to be an advantage for males because it prevents the production of offspring of other males. During copulation, a male transfers seminal fluid containing ACPS protein to a female, and ACPS protein influences the metabolism and physiology of females

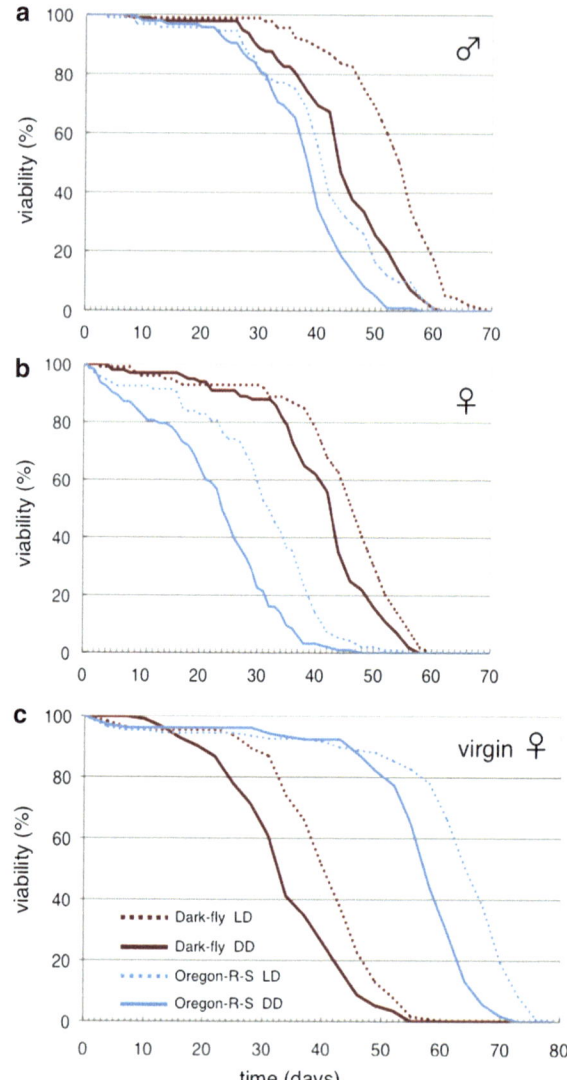

Fig. 4.4 Survival curves of Dark-fly and Oregon-R-S. The viability of male flies (**a**) and female flies (**b**) reared together is plotted versus time (days). Dark-fly (*dark brown lines*) and Oregon-R-S (*light blue lines*) was reared under LD (*dotted lines*) or DD (*solid lines*) conditions. The viability of virgin females (**c**) was also measured in a similar manner. $n = 92$–100 flies. Oregon-R-S virgin females showed longer longevity than the mated ones, whereas Dark-fly virgin females showed shorter longevity than the mated ones

(Ravi Ram and Wolfner 2007). It has also been proposed that some volatile compounds emanated from males cause deleterious effects on females in the absence of mating (Partridge and Fowler 1990). We speculate that Dark-fly females might be resistant to such deleterious compounds, and that Oregon-R-S females might be sensitive to them, especially in the dark. Alternatively, these phenomena might be due to the traits of males; for example, seminal fluid of Dark-fly males might not be deleterious to females.

Taken together, our experiments revealed that Dark-fly produced more offspring in the dark than in the light, and Dark-fly females survived longer than Oregon-R-S females, especially in the dark. These traits of Dark-fly would contribute to reproductive success that would potentially be adaptive for living in the dark.

4.3   Genome of Dark-fly

4.3.1   Whole Genome Sequencing for Dark-fly

To understand the molecular nature of Dark-fly's traits, we extracted genomic DNA from Dark-fly and Oregon-R-S, and performed whole genome sequencing for them using a next generation sequencer (Illumina Genome Analyzer II) (Fig. 4.5a). Approximately 67 million and 87 million reads (i.e. identified nucleotide sequences) were obtained for Dark-fly and Oregon-R-S, respectively, and 96 % and 90 % of these reads were successfully aligned to the *Drosophila* reference genome (Table 4.1). Since the read sequence for Dark-fly covered the genome with mean depth of 14, our data were suitable for comprehensively analyzing the features of the genome. After filtering the quality of each sequence, SNPs were identified at 415,626 sites for Dark-fly and 415,668 sites for Oregon-R-S, compared with the reference genome sequence (Table 4.2). 198,286 SNPs (47.7 % for Dark-fly) were shared between the two lines, and 217,340 SNPs were specifically identified in Dark-fly (Fig. 4.6a).

Although Dark-fly was derived from the Oregon-R-S strain, the genome sequences of the present Dark-fly and the present Oregon-R-S were thus found to be somewhat divergent. Previous studies evaluated the spontaneous nucleotide mutation rate in *Drosophila* and estimated it to be $1/10^9$–$1/10^8$ per nucleotide per generation (Keightley et al. 2009; Haag-Liautard et al. 2007), which is a value that is approximately conserved among diverse organisms (Denver et al. 2004). Given that most newly arisen mutations have been fixed in a relatively small population (about 100 flies) of Dark-fly, we estimated that 400–4,000 mutations would arise during 1,400 generations by the following simple calculation: mutation rate $(1/10^9$–$1/10^8) \times$ genome size $(1.5 \times 10^8$ bases $\times 2) \times$ generations (1,400 generations). Therefore, the number of SNPs found between Dark-fly and Oregon-R-S would be 55–550 times greater than the predicted number, if two lines had been derived from exactly the same ancestor. This discrepancy might be explained by several possibilities. One possibility is that Oregon-R-S strains might have originally been diverse in the stocks in different laboratories. Another possibility is that the mutation rate in one of the strains was accelerated, for example via mutation in a DNA polymerase enzyme (Barrick et al. 2009). Alternatively, unexpected

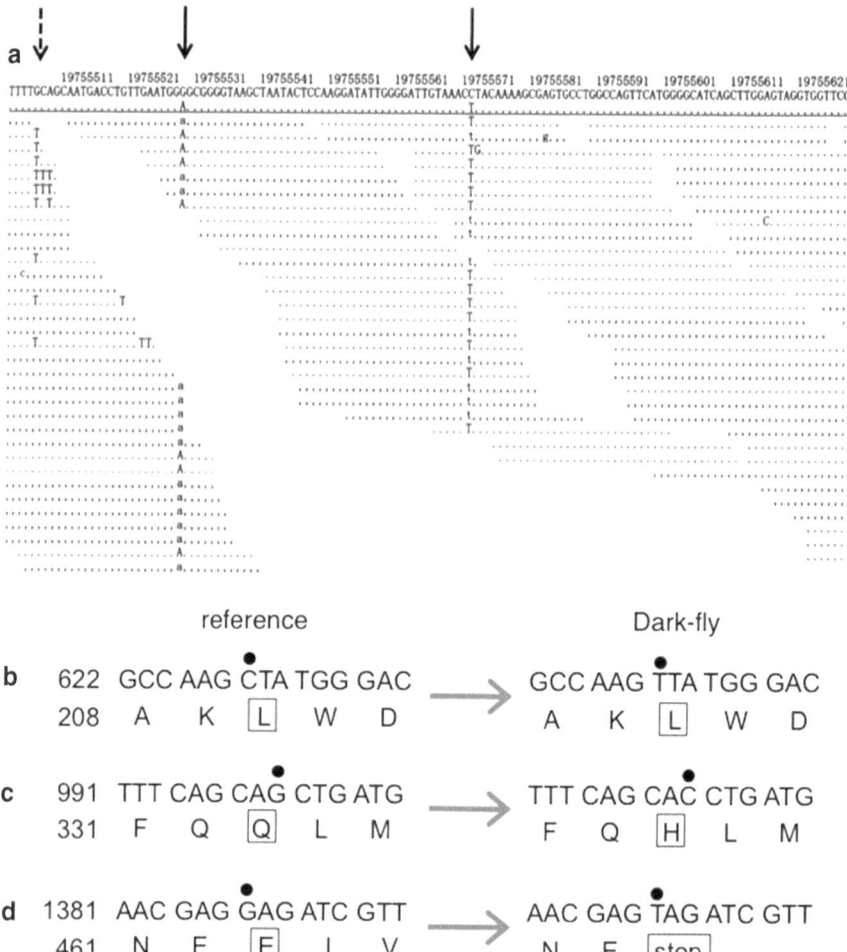

Fig. 4.5 Whole-genome sequencing for Dark-fly. (**a**) An example of NGS data is shown. The numbers and the sequence at the top indicate the nucleotide positions on the 2R chromosome and the reference genome sequence. The numerous *dotted lines* show the aligned "reads" (each 36 nucleotides long) from the sequence data of Dark-fly. *Dots* mean the same sequence as the reference sequence, while alphabetical characters indicate a different sequence. Two *arrows* and one *dotted arrow* show homozygous (an altered nucleotide is shared by all reads) and heterozygous (an altered nucleotide is partially shared for reads) SNPs, respectively. (**b**) An example of a synonymous SNP in the Gbeta13F gene. An altered nucleotide (marked by a *dot*) does not change the amino acid sequence. Upper and lower sequences are nucleotide and amino acid sequences, respectively. Numbers on the left represent the nucleotide number of the coding region and amino acid number of the protein. (**c**) An example of a nonsynonymous SNP in the alpha-Esterase6 gene. An altered nucleotide causes an amino acid substitution (from Q to H). (**d**) An example of a nonsense mutation in the Rh7 gene. An altered nucleotide produces a premature stop codon in the amino acid sequence

Table 4.1 Summary of genome sequencing

Fly line	Read length	Read number	Mapped read %	Mean depth
Dark-fly	36	66,855,594	96.4	13.7
Oregon-R-S	36, 39, 48	87,101,330	89.7	19.6

The results of genome sequencing using an Illumina Genome Analyzer II are summarized. Flybase Dmel 5.22 genome (168,736,537 bases) was used as a reference genome

Table 4.2 SNP and InDel analyses

Fly line	Dark-fly	Oregon-R-S
Total fixed SNPs	415,626	415,668
SNP frequency (bases/SNP)	406	406
Line-specific SNPs	217,340	217,382
Line-specific nsSNPs	9,695	6,521
Genes carrying nsSNPs	4,323	3,039
Genes carrying nonsense mutations	28	23
Total fixed InDels	5,322	5,461
InDel frequency (bases/InDel)	31,705	30,898
Line-specific InDels	4,660	4,799
Line-specific cInDels	52	27
Genes carrying cInDels	50	27

These data represent a summary of our analyses of SNPs and InDels for the Dark-fly and Oregon-R-S genomes

Fig. 4.6 Analysis of SNPs in the Dark-fly genome. (**a**) 415,626 SNPs were identified for Dark-fly, and 198,286 SNPs (47.7 % of the number in Dark-fly) were shared with Oregon-R-S. (**b**) Dark-fly SNPs were classified into 1,435,028 SNP-effects. The proportion of SNP-effects (intergenic, UTR and intron of mRNA, coding synonymous, and coding nonsynonymous) is shown

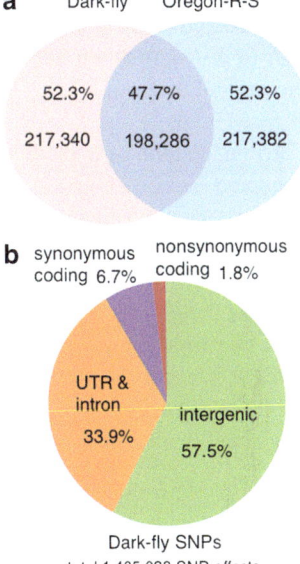

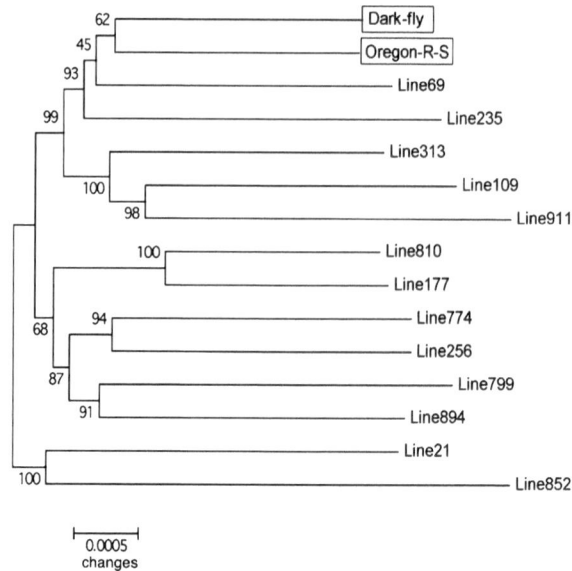

Fig. 4.7 Phylogenetic tree analysis of the Dark-fly and Oregon-R-S genomes. Neighbor-joining tree with p-distance based on analysis of the combined nucleotide sequence (total 87,982 bases) of eight genes (aru, chic, betaInt-nu, Khc-73, insc, drpr, glec, and tau). The bootstrap value from 500 replications was calculated using MEGA software. DGRP lines were randomly selected and are indicated by line number. DGRP lines were highly diverse, whereas Dark-fly and Oregon-R-S were relatively close on the phylogenetic tree

contamination might have occurred during the history of the strains. It is impossible to distinguish among these possibilities at present, because we have neither the original fly from 57 years ago nor sister lines maintained in parallel with Dark-fly (see Chap. 1).

To understand how close the genome of Dark-fly is to that of Oregon-R-S, we compared them with genomes of a group of other lines (the DGRP lines) (Mackay et al. 2012), which are inbred lines generated from a natural population, as described above. Phylogenetic tree analysis revealed that the DGRP lines are highly diverse, whereas Dark-fly and Oregon-R-S are relatively close (Fig. 4.7), suggesting that although the present Dark-fly has many SNPs compared to the present Oregon-R-S, these two lines are closely related. We also examined the mitochondrial genome, which is maternally inherited and is not subject to recombination. Twelve of 16 SNPs (75 %) found in Dark-fly corresponded to those of Oregon-R-S (12 of 19), suggesting that the maternal origins of the two lines were related.

4.3.2   Analysis of Dark-fly SNPs and InDels

Since Dark-fly displays some traits advantageous for living in the dark, it should carry some genomic alterations related to these traits. Even if so, most of the SNPs we found would be expected to be functionally neutral and only a small fraction of the SNPs should contribute to the adaptive traits. To evaluate the Dark-fly SNPs, we categorized each SNP by its position relative to gene structures, such as intergenic regions and gene coding regions. Since one SNP often affects several isoforms of a gene or several overlapping genes simultaneously, the 415,626 SNPs of Dark-fly were classified into 1,435,028 SNP-effects (Fig. 4.6b). Among them, 57.5 % were mapped to intergenic regions, and 33.9 % were mapped to untranslated regions (UTRs) or introns of transcribed regions. Only 8.5 % of the SNPs were located within protein coding regions of genes. It is not easy to evaluate SNPs in non-coding regions, and accordingly we focused on the coding SNPs. 6.7 % of the SNP-effects were synonymous SNPs (sSNPs: i.e., they do not alter amino acid sequences of gene products) (Fig. 4.5b), and 1.8 % were non-synonymous SNPs (nsSNPs: i.e., they change the amino acid sequence) (Fig. 4.5c). We collected the Dark-fly-specific nsSNPs without redundancy between isoforms and identified 4,323 genes carrying nsSNPs. Since nsSNPs could alter the activity of gene products, these genes carrying nsSNPs would be primary candidates contributing to the altered traits of Dark-fly (Table 4.2). We performed similar processes for the Oregon-R-S genome and identified 3,039 such genes.

An InDel is an insertion or deletion of a few nucleotides and can be detected by analyzing NGS data. We identified 5,322 and 5,461 InDels for Dark-fly and Oregon-R-S, respectively, and 662 of these InDels (12.4 % for Dark-fly) were shared between them (Table 4.2). We classified each InDel by its position relative to gene structures, by a process similar to that performed for SNP analysis. InDels in gene coding regions (cInDels) would result in codon-deletion, codon-insertion, or frame-shift of gene products, so that the effects of cInDels would be severe, like those of nsSNPs. We identified 50 and 27 cInDels specifically found in Dark-fly and Oregon-R-S, respectively (Table 4.2).

We then asked whether the nsSNP or cInDel-carrying genes are concentrated in any gene families in the Dark-fly genome. Gene Ontology (GO) family is a useful criterion to classify genes according to molecular and biological functions of gene products (Huang da et al. 2009). We identified nine GO families that contained nsSNPs or cInDels at higher probability than the average for all genes throughout the genome (Table 4.3). Among them, four GO families, including families associated with metal ion binding (GO:0046872) and UDP-glycosyltransferase activity (GO:0008194), were shared between Dark-fly and Oregon-R-S (* in Table 4.3), suggesting that these genes might have been commonly subject to mutations. The remaining five GO families were found specifically for Dark-fly (Table 4.3). These include families associated with carboxylesterase activity (GO:0004091) and guanyl-nucleotide exchange factor activity (GO:0005085). Thus, these gene families have accumulated nsSNPs and cInDels in the Dark-fly genome and might be potentially related to the Dark-fly's traits.

Table 4.3 GO families of genes carrying nsSNPs and cInDels in Dark-fly

GO term: molecular function *: shared with Oregon-R-S	Total gene #	Count#
* GO:0046872 ~ metal ion binding	1,718	609
* GO:0004888 ~ transmembrane receptor activity	383	159
* GO:0008194 ~ UDP-glycosyltransferase activity	92	45
GO:0003700 ~ transcription factor activity	389	151
GO:0008061 ~ chitin binding	100	47
* GO:0016758 ~ transferase activity, transferring hexosyl groups	138	61
GO:0004091 ~ carboxylesterase activity	107	47
GO:0005085 ~ guanyl-nucleotide exchange factor activity	60	29
GO:0043565 ~ sequence-specific DNA binding	230	89

GO families (molecular function) of genes carrying nsSNPs and cInDels at high probability (p-value <0.01) were listed for Dark-fly using the DAVID tool. * indicates a family shared with that listed for Oregon-R-S (data not shown). Total gene number annotated in each GO family and gene number counted from the Dark-fly data are indicated

4.3.3 Nonsense Mutations in the Dark-fly Genome

Among nsSNPs, a nonsense mutation produces a stop codon in the amino acid sequence of a gene product, and may severely affect the protein's function (Fig. 4.5d). We identified 28 nonsense mutations in the Dark-fly genome (Table 4.4). Among them, ten mutations (for example, in the Hn and HisCl1 genes) were located in a subset of a gene's isoforms, so that the nonsense mutation might be complemented by redundant function(s) of other isoform(s). The remaining 18 mutations were located at sites shared by all of a gene's isoforms or at sites of a gene encoding a unique transcript, so that functional consequences of these mutations would be inevitable. These genes included an olfactory receptor (Or65c) and a light receptor (Rh7) gene.

It has been proposed that olfactory receptor genes evolve rapidly in a non-neutral manner, and often become pseudogenes (non-functional genes) (Gardiner et al. 2008). According to this notion, mutations of these genes would generate diversity of odor discrimination between species and even between individuals. In the Dark-fly genome, we detected nsSNPs in 36 of 59 olfactory receptor (Or) genes, in addition to the nonsense mutation in the Or65c gene. These mutations might be related to odor discrimination of Dark-fly.

Rhodopsin is a light-sensing receptor that belongs to the G protein-coupled receptor family, and the *Drosophila* genome encodes seven rhodopsins (Wang and Montell 2007; Katz and Minke 2009). The Dark-fly genome contains a nonsense mutation in the rhodopsin7 (Rh7) gene but no nsSNPs in other rhodopsin genes. Although the in vivo functions of Rh7 are still unclear, it is known that the Rh7 protein possesses a unique structure: both its N- and C-terminal regions are longer and its third cytoplasmic loop is shorter than those of other rhodopsins. A nonsense

Table 4.4 Genes carrying nonsense mutations in the Dark-fly genome

Gene name	GO: molecular function	Iso forms
CG3332	–	Sub
CG7236	Cyclin-dependent protein kinase activity	Sub
CG11321	Zinc ion binding	Sub
CG13124	–	Sub
CG15260	–	All
CG18478	Serine-type endopeptidase activity	All
CG31782	Nucleic acid binding	All
CG31792	Transmembrane movement of substances	All
CG17493	Calcium ion binding	All
CG7882	Glucose transmembrane transporter	All
CG18538	–	All
CG5549	Glycine: sodium symporter activity	All
CG14950	–	All
Or65c	Olfactory receptor activity	All
Hn	Phenylalanine 4-monooxygenase activity	Sub
CG6745	Pseudouridine synthase activity	All
CG6776	Glutathione transferase activity	All
Cpr67Fa1	Chitin-based larval cuticle	All
Rh7	G-protein coupled photoreceptor activity	All
Or69a	Olfactory receptor activity	Sub
CG9795	–	Sub
HisCl1	Histamine-gated chloride channel	Sub
Rpb4	Histone acetyltransferase activity	Sub
CG10183	Transferring acyl groups	All
CG33658	–	All
CG4774	CDP-alcohol phosphatidyltransferase	All
Vha100-1	Hydrogen-exporting ATPase activity	Sub
CG2854	–	All

Twenty-eight nonsense mutations were identified in the Dark-fly genome. Genes carrying nonsense mutations and GO terms of genes are shown. Ten mutations were located in a subset of gene isoforms (indicated as "sub" in the "isoforms" column), and 18 mutations altered all products of a gene (indicated as "all")

mutation in Dark-fly is located in the C-terminal region and results in the truncation of 21 amino acids from the C-terminus of the wild-type Rh7 protein (483 amino acids long) (Fig. 4.5d). We suggest that the long C-terminal region plays some roles in the functions of Rh7 because the entire amino acid sequence of the Rh7 protein is highly conserved between the *Drosophila* genus and some other insects. An interesting issue for future research is whether Dark-fly lost the function of a useless light receptor or whether the truncated receptor gained a new function in Dark-fly. Even more importantly, we need to know the in vivo function of Rh7 in the wild-type fly.

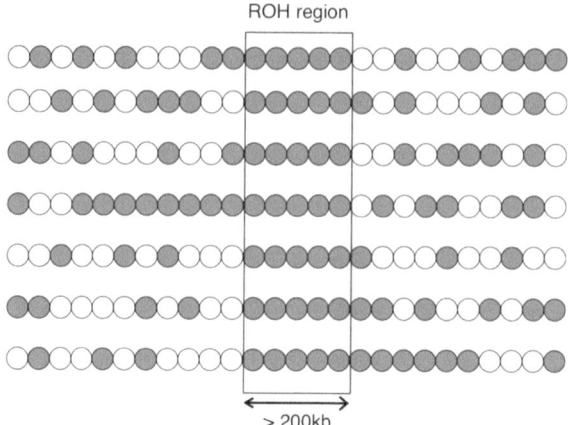

Fig. 4.8 Run of homozygosity region. A schematic drawing of the population genome. Each *chain of circles* represents an individual genome with *white* and *grey circles* showing reference sequences (non SNPs) and SNPs, respectively. An ROH region is a genomic region (more than a few hundred kb) containing consecutive homozygous SNPs shared by all individuals

4.3.4 Genome Regions Selected in Dark-fly

To identify the genome regions selected during the history of Dark-fly, we characterized the genome structure of the Dark-fly population. Runs of homozygosity (ROH) regions are homozygosity-extended genomic regions (more than a few hundred kb) containing consecutive homozygous SNPs, and are thought to be regions currently selected in a population's genome (Fig. 4.8) (McQuillan et al. 2008). This criterion has successfully identified disease-related recessive mutations and positively selected genes in human populations (Evans et al. 2005; Lencz et al. 2007). We expected that the Dark-fly genome might contain homozygosity-extended regions as signatures of historical selections during its 1,400 generations in the dark. Since our NGS data were obtained from the genomic DNA of 20 male flies and cover the genome with 14-fold depth, we considered that our data would be useful to detect ROH regions in the population genome. We listed homozygous SNPs (homo SNPs; frequency greater than 90 %) and heterozygous SNPs (hetero SNPs; frequency greater than 40 % and less than 90 %) from the Dark-fly genome data (for example, Fig. 4.5a arrow and dotted arrows) and identified 449,684 homo SNPs and 28,132 hetero SNPs (Table 4.5). The overall fraction of homo SNPs was 94.1 %, indicating that the Dark-fly genome contains only a small number of hetero SNPs compared to homo SNPs. We searched homozygosity-extended regions (400-kb sliding window with 200-kb steps) on major chromosomes (2L, 2R, 3L, 3R and X) and identified 24 ROH regions (Table 4.5). The total length of ROH regions covered approximately 6 Mb (5 % of the genome length of major chromosomes), suggesting that homo SNPs are abundant but ROHs are rare in the Dark-fly genome. We performed a similar process for Oregon-R-S and identified 128 ROH regions that

Table 4.5 Identification of ROH regions

Fly line	Dark-fly	Oregon-R-S
Homo and hetero SNPs (0.4 = <freq)	477,816	486,013
Homo SNPs (0.9 = <freq)	449,684	453,646
Hetero SNPs (0.4 = <freq < 0.9)	28,132	32,367
Homo SNP fraction in total (%)	94.1	93.3
Number of ROH regions	24	128
Total length of ROHs (kb)	5,934	43,868
Fraction of ROHs in genome (%)	4.99	36.85
Number of ROH regions with significantly high homozygosity	21	ND
Genes carrying nsSNPs and cInDels in ROH regions	241	ND

These data present a summary of our analyses of ROH regions. Homo and hetero SNPs were identified. The number of homo SNPs was slightly different from that of the fixed SNPs (Table 4.2), due to the difference of data filtering. ROH regions were identified using PLINK software. The Dark-fly ROH regions showing significantly high homozygosity were determined by statistical analyses. Genes carrying nsSNPs and cInDels in 21 ROH regions were counted. ND means not determined

covered approximately 44 Mb (37 % of the genome length of major chromosomes) (Table 4.5). Thus, although the percentages of homo SNPs were similar between Dark-fly and Oregon-R-S (94.1 % versus 93.3 %), the ROH number and coverage were clearly different between them. This indicates that homo and hetero SNPs are highly clustered in the Oregon-R-S genome but are distributed more evenly in the Dark-fly genome, resulting in the presence of many ROHs in Oregon-R-S and few ROHs in Dark-fly. These genome features might reflect the differences of population history. For example, inbreeding (isogenization) might have occurred frequently for Oregon-R-S during its history, and consequently many SNPs might have become fixed as clusters in the population genome. In contrast, Dark-fly has been maintained mostly as a constant population size (about 100 flies), and many genomic regions might still be fluctuating in the population. If this is true, it would strongly support the notion that the Dark-fly ROH regions are rare genome regions selected during the current history (57 years).

We also measured mean homozygosity (mean frequency of each SNP) in the Dark-fly and Oregon-R-S genomes. Our analysis revealed that in both lines, high homozygosity was expanded widely throughout the genome and only a small number of regions showed low homozygosity (Fig. 4.9). This seems to be a genome feature of inbred organisms. In most genomic regions, the Oregon-R-S genome displayed higher homozygosity than the Dark-fly genome, consistent with the difference of ROH number and coverage (see Fig. 4.9, light blue and dark brown lines). We evaluated each ROH region of the Dark-fly genome by statistical tests and finally identified 21 ROH regions showing significantly high homozygosity (Table 4.5 and Fig. 4.9). We suggest that these ROH regions might be genome signatures selected in the Dark-fly population.

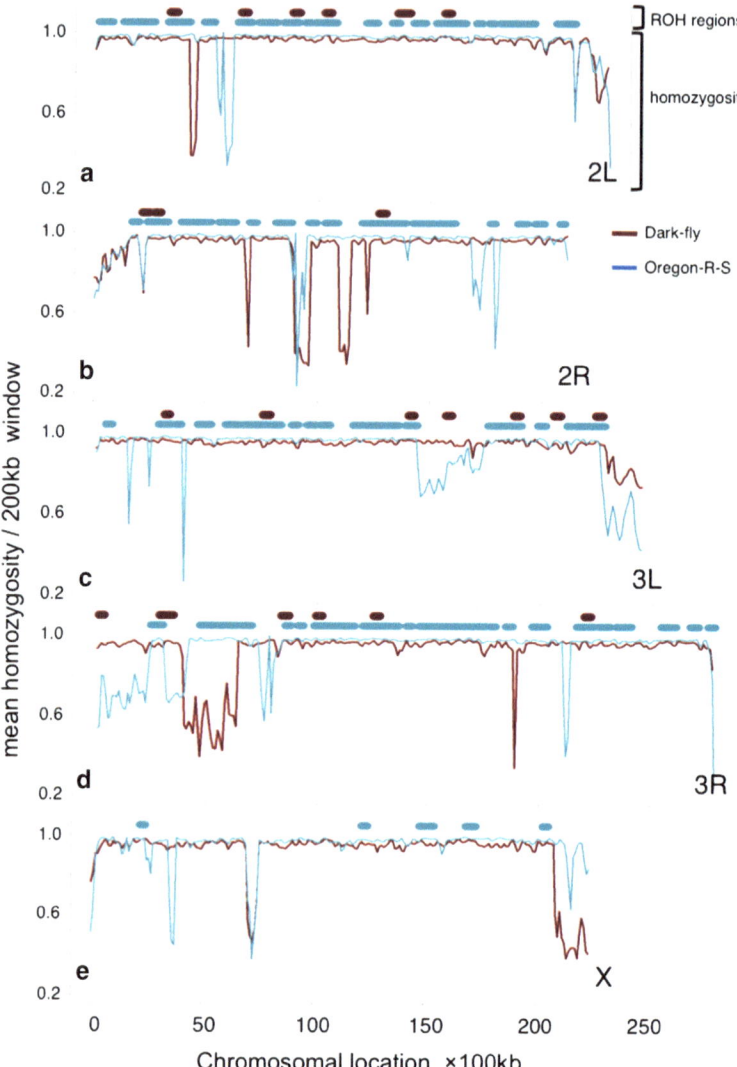

Fig. 4.9 Homozygosity and ROH regions. Mean homozygosity of SNPs in a sliding window (200-kb window at 100-kb steps) was plotted versus the location on the 2L (**a**), 2R (**b**), 3L (**c**), 3R (**d**) and X (**e**) chromosomes. The Oregon-R-S genome (*light blue lines*) displayed higher homozygosity than the Dark-fly genome (*dark brown lines*) in most regions. *Thick horizontal bars* represent ROH regions identified by PLINK software for Oregon-R-S (*light blue bars*) and Dark-fly (*dark brown bars*) and are plotted above the graph without homozygosity values

Table 4.6 Genes carrying nsSNPs and cInDels in the Dark-fly ROH regions

ROH ID#	Genes carrying nsSNPs or cInDels
ROH1	CG8838, CG34394, Ptpa, CG34175, CG31952, CG3238, CG31776, Sr-CIV, Spindly
ROH2	CG9596, CG11319, CG11320, CG34345, Oatp26F, Tango1, CG31633, CG11070, CG13771, Nhe3, CG11327, GRHR, CG11188, homer, TTLL3A, CG31910, CG11221, CG11322, CG11321, CG17378
ROH3	CG32986, CG34398, CG9510, CG31886, CG32985, CG32984, CG18088, CG9541, CG9555, CG17906, CG18661, CG9568, CG9582, Toll-4
ROH4	CG34043, CG5604, CG13138, CG5384, CG4972, GATAd, CG34367, CG5367, Cand1, pim, CG5056, rho-5, CG33303, gny, CG5168, CG5188, CG6232, CG5322, CG6206, RluA-1, RluA-2, CG7456, CG13144, Myo31DF, CG7384, Fatp
ROH5	CG33641, CG33644, CG33645, CG16853, CG18507, CG7311, CG31814, CG9014, CR31845, CG31731, **sec71**
ROH6	CG16865, **Sos**, b, tam, Orc5, mRpS23, CG33307, CG33306, CG8997, cenG1A, Ance-2, CG16886, CG16884, nimB1, nimB3, nimB5, He, nimC1, rk, bgm, CG18095
ROH7	CG7631, CG18480, CG4587, CycE, Ku80, CG18109, CG18518
ROH9	CG15236, Spn42Db, Spn42De, CG3358, CheB42b, CheB42c, ppk25, mim, Cyp6u1, CG30157, vimar, Tsp42Ee, Tsp42Eh, Tsp42Ei, CG12831
ROH10	Fen1, CG8910, Pkc53E, CG15614, mute, CG6665, ste24b, CG6796, CG8963, Ark, **RhoGEF2**, CG9640, CG9642, CG9646, CG8950, CG6967, CG30460, **CG30456, CG15611**
ROH11	CG14963, CG32284, CG32277, CG12034, CG11505, CG12009
ROH13	pex1, CG8100, Fbp1, Sox21b, nuf, CG34244
ROH14	CG13445, CG12713, CG32150, CG12486, pHCl, sff, Pka-C3
ROH15	CG14073, CR32027, CG14074, dysb, CG11637, Ir75d, CG14077, CG3819, CG14075, CG11619, CG18135, CG3808, CG18136, nkd
ROH16	CG13251, CG34260, CG13252, CG4074, Pitslre, Spc105R
ROH17	CG14459, CG14453, CG11370, CG6838, CG32454
ROH19	CG1988, CG1105, CG1965, CG1943, CG1091, CG31248, MAGE, lap, CG14605, CG1227
ROH20	CG14598, **alpha-Est10, alpha-Est9, alpha-Est8, alpha-Est7, alpha-Est6, alpha-Est5, alpha-Est3, alpha-Est2, CG34127**
ROH21	Octbeta2R, CG11608, Cyp313a4, CG14391, mus308, Men, CG5724, CG5999
ROH22	CCHa1, Or88a, Kif19A, 140up, CG14356, CG42500, CG31533, CG31327, DopR, CG9649, CG9631, Aats-met, trx, CG3259, su(Hw), CG31321
ROH23	Ubx, Glut3, Abd-B
ROH24	CG14239, Hex-t1, CG5455, CG6490

Twenty-one Dark-fly ROH regions showing significantly high homozygosity are listed. Genes carrying nsSNPs and InDels in each ROH region are shown. Five GEF genes and nine esterase genes are shown as bold characters

4.3.5 SNPs and InDels in the Selected Genome Regions

We further characterized the Dark-fly ROH regions and identified 241 genes containing nsSNPs and/or cInDels (Table 4.6). GO analysis for the 241 genes listed three families. One of them is associated with carboxylesterase activity (GO number: 0004091), and the others are related families associated with guanyl-nucleotide exchange factor activity (GO number: 0005085). Interestingly, both families were also listed by the aforementioned GO analysis of total nsSNPs and cInDels (Table 4.3). Carboxylesterase is a family of enzymes hydrolyzing esters, and the alpha-esterase class listed here is involved in xenobiotic metabolism (Claudianos et al. 2006). Guanyl-nucleotide exchange factors (GEFs) are regulators of small GTPases involved in various biological processes, such as neural development and activity (Schmidt and Hall 2002). These and other genes that carry nsSNPs and cInDels in the ROH regions are potential candidate genes related to the selected traits of Dark-fly (Table 4.6).

Alpha-esterases are involved in the metabolism of xenobiotics (so-called detoxification) (Claudianos et al. 2006). Although the targets of each alpha-esterase are still unclear, some alpha-esterases function in resistance against pesticides, such as organophosphates (Heidari et al. 2005). Interestingly, GO analysis of total nsSNPs and cInDels listed another gene family related to detoxification, namely UDP-glycosyltransferase (UGT) genes (Luque and O'Reilly 2002), as well as the esterase family. The UGT family was listed for both Oregon-R-S and Dark-fly, though the mutation rate in this gene family was higher in Dark-fly. Thus, Dark-fly nsSNPs and cInDels are concentrated in two detoxification enzyme families. It is known that alpha-esterase and UGT genes are expressed under circadian regulation in *Drosophila* as well as in other animals (Claridge-Chang et al. 2001). Indeed, flies' resistance against pesticides oscillates daily (Hooven et al. 2009). Although a previous study showed that locomotor activity of Dark-fly displays normal circadian rhythm (see Chap. 2) (Imafuku and Haramura 2011), the intriguing question of whether the detoxification rhythm is changed in Dark-fly has not yet been answered. The biological meaning of detoxification rhythms is still mysterious, but they are expected to promote cost-effective performance during feeding time, when flies are exposed to chemical compounds from the environment. We also speculate that light itself might influence the detoxification process. It is known that bilirubin, a human xenobiotic derived from heme, is metabolized by UGT and that light exposure bypasses the requirement for UGT in this process (Beutler et al. 1998). We note another example, namely, that active vitamin D is synthesized in human skin upon exposure to UV light and is metabolized through the action of a detoxification enzyme, cytochrome P-450 (Zhu and DeLuca 2012). Thus, light energy chemically alters multiple compounds associated with xenobiotic metabolism pathways. One hypothesis is that the normal fly might need light to metabolize xenobiotics effectively, and therefore the absence of light would affect the fly's physiology and behavior (Fig. 4.10). Dark-fly might possess specialized metabolism of xenobiotics in light-free conditions. We need to test this hypothesis in the future. It is also known that some vertebrate detoxification enzymes are preferentially expressed in olfactory epithelium and act in the clearance of odors after perception (Lazard et al. 1991). Similarly, some *Drosophila* enzymes are specifically expressed in the olfactory organ (Wang et al. 1999). We speculate that the detoxification enzymes might be related to olfactory ability in Dark-fly.

The Dark-fly ROH regions also contain five guanyl-nucleotide exchange factor (GEF) genes carrying nsSNPs and cInDels. GEFs are regulators of small GTPases involved in various biological processes, such as neural development and activity.

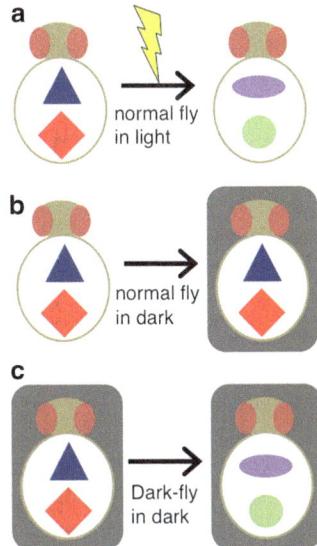

Fig. 4.10 Hypothetical model of Dark-fly's detoxification. (**a**) Light energy converts multiple compounds (illustrated as conversion from *triangle* and *lozenge* to *ellipse* and *circle*). Normal fly might need light to metabolize xenobiotics effectively. (**b**) Normal fly in the dark might contain residual xenobiotics, and this might affect the fly's physiology and behavior. (**c**) Dark-fly carries many mutations in detoxification enzyme genes, and thus might possess specialized metabolism of xenobiotics in the dark

For example, the Sos gene is required for the development of R7 photoreceptor neurons (Rogge et al. 1991) and is also involved in circadian rhythms of clock neurons (Williams et al. 2001). RhoGEF2 protein shapes the morphology of cells and functions in axonal growth (Ng and Luo 2004). Recently, Yuan et al. found that the morphology of larval photoreceptor neurons is plastically changed by light and dark conditions (Yuan et al. 2011). An intriguing issue for future studies is whether Dark-fly retains this neural plasticity.

4.4 Conclusions and Future of Dark-fly

We determined the whole genome sequence of Dark-fly using NGS technology and identified many SNPs and InDels. By analyses of population genome structure, we detected genome regions likely selected in the Dark-fly's history, and found that nsSNPs and cInDels were preferentially accumulated in some gene families. We also identified some nonsense mutations, which would disrupt or severely affect the function of the gene products. These data provided a list of potential candidate genes related to Dark-fly's traits. These candidate genes include a light receptor, olfactory receptors, and enzymes related to detoxification and neural development. Some of these genes might contribute to gain of useful traits or loss of useless traits in the dark environment. Alternatively, some of these genes might contribute to trade-offs between useful traits and useless traits, as demonstrated in cavefish (Yamamoto et al. 2009). Further analyses of candidate genes will clarify the effects of these mutations in Dark-fly. Since we evaluated SNPs and InDels using limited criteria, we have not excluded the possibility that other (coding and noncoding) mutations not discussed here contribute to the environmental adaptation. Also, since Dark-fly has been reared with a minimal medium, it is possible that Dark-fly might be adapted to poor nutrients as well as to the dark, and the genomic alterations we found might be related to the adaptation to the nutrient state.

The whole-genome sequencing described here is a first step toward linking genome, trait and adaptation. As a second step, we are now maintaining large mixed populations of Dark-fly and Oregon-R-S in different conditions and will examine the dark-selected SNPs in the population genome. Another intriguing issue yet to be resolved is whether Dark-fly has an altered profile of gene expression. NGS technology will be useful for experiments to address this issue, and will provide us a wide array of approaches for experimental evolution studies.

References

Ashburner M, Golic KG, Hawley SR (2005) *Drosophila*: a laboratory handbook. Cold Spring Harbor Laboratory, New York

Barrick JE, Yu DS, Yoon SH, Jeong H, Oh TK, Schneider D, Lenski RE, Kim JF (2009) Genome evolution and adaptation in a long-term experiment with *Escherichia coli*. Nature 461:1243–1247. doi:10.1038/nature08480

Beutler E, Gelbart T, Demina A (1998) Racial variability in the UDP-glucuronosyltransferase 1 (UGT1A1) promoter: a balanced polymorphism for regulation of bilirubin metabolism? Proc Natl Acad Sci USA 95:8170–8174

Billeter J-C, Rideout EJ, Dornan AJ, Goodwin SF (2006) Control of male sexual behavior in *Drosophila* by the sex determination pathway. Curr Biol 16:R766–R776. doi:10.1016/j.cub.2006.08.025

Burke MK, Dunham JP, Shahrestani P, Thornton KR, Rose MR, Long AD (2010) Genome-wide analysis of a long-term evolution experiment with *Drosophila*. Nature 467:587–590. doi:10.1038/nature09352

Chou H-H, Chiu H-C, Delaney NF, Segrè D, Marx CJ (2011) Diminishing returns epistasis among beneficial mutations decelerates adaptation. Science 332:1190–1192. doi:10.1126/science.1203799

Claridge-Chang A, Wijnen H, Naef F, Boothroyd C, Rajewsky N, Young MW (2001) Circadian regulation of gene expression systems in the *Drosophila* head. Neuron 32:657–671. doi:10.1016/s0896-6273(01)00515-3

Claudianos C, Ranson H, Johnson RM, Biswas S, Schuler MA, Berenbaum MR, Feyereisen R, Oakeshott JG (2006) A deficit of detoxification enzymes: pesticide sensitivity and environmental response in the honeybee. Insect Mol Biol 15:615–636

da Huang W, Sherman BT, Lempicki RA (2009) Systematic and integrative analysis of large gene lists using DAVID bioinformatics resources. Nat Protoc 4:44–57. doi:10.1038/nprot.2008.211

Darwin C (1859) On the origin of species. Murray, London

Denver DR, Morris K, Lynch M, Thomas WK (2004) High mutation rate and predominance of insertions in the *Caenorhabditis elegans* nuclear genome. Nature 430:679–682

Edrey YH, Park TJ, Kang H, Biney A, Buffenstein R (2011) Endocrine function and neurobiology of the longest-living rodent, the naked mole-rat. Exp Gerontol 46:116–123. doi:10.1016/j.exger.2010.09.005

Evans PD, Gilbert SL, Mekel-Bobrov N, Vallender EJ, Anderson JR, Vaez-Azizi LM, Tishkoff SA, Hudson RR, Lahn BT (2005) Microcephalin, a gene regulating brain size, continues to evolve adaptively in humans. Science 309:1717–1720. doi:10.1126/science.1113722

Gardiner A, Barker D, Butlin RK, Jordan WC, Ritchie MG (2008) *Drosophila* chemoreceptor gene evolution: selection, specialization and genome size. Mol Ecol 17:1648–1657

Greenspan RJ, Ferveur JF (2000) Courtship in *Drosophila*. Annu Rev Genet 34:205–232. doi:10.1146/annurev.genet.34.1.205

Gross JB, Protas M, Conrad M, Scheid PE, Vidal O, Jeffery WR, Borowsky R, Tabin CJ (2008) Synteny and candidate gene prediction using an anchored linkage map of *Astyanax mexicanus*. Proc Natl Acad Sci USA 105:20106–20111. doi:10.1073/pnas.0806238105

Gross JB, Borowsky R, Tabin CJ (2009) A novel role for Mc1r in the parallel evolution of depigmentation in independent populations of the cavefish *Astyanax mexicanus*. PLoS One 5:e1000326

Gustafsson MV, Zheng X, Pereira T, Gradin K, Jin S, Lundkvist J, Ruas JL, Poellinger L, Lendahl U, Bondesson M (2005) Hypoxia requires Notch signaling to maintain the undifferentiated cell state. Dev Cell 9:617–628. doi:10.1016/j.devcel.2005.09.010

Haag-Liautard C, Dorris M, Maside X, Macaskill S, Halligan DL, Houle D, Charlesworth B, Keightley PD (2007) Direct estimation of per nucleotide and genomic deleterious mutation rates in *Drosophila*. Nature 445:82–85. doi:10.1038/nature05388

Heidari R, Devonshire AL, Campbell BE, Dorrian SJ, Oakeshott JG, Russell RJ (2005) Hydrolysis of pyrethroids by carboxylesterases from *Lucilia cuprina* and *Drosophila melanogaster* with active sites modified by in vitro mutagenesis. Insect Biochem Mol Biol 35:597–609. doi:10.1016/j.ibmb.2005.02.018

Hermisson J, Pennings PS (2005) Soft sweeps. Genetics 169:2335–2352. doi:10.1534/genetics.104.036947

Hooven LA, Sherman KA, Butcher S, Giebultowicz JM (2009) Does the clock make the poison? Circadian variation in response to pesticides. PLoS One 4:e6469. doi:10.1371/journal.pone.0006469

Imafuku M, Haramura T (2011) Activity rhythm of *Drosophila* kept in complete darkness for 1300 generations. Zoolog Sci 28:195–198. doi:10.2108/zsj.28.195

Izutsu M, Zhou J, Sugiyama Y, Nishimura O, Aizu T, Toyoda A, Fujiyama A, Agata K, Fuse N (2012) Genome features of "dark-fly", a *Drosophila* line reared long-term in a dark environment. PLoS One 7:e33288. doi:10.1371/journal.pone.0033288

Jeffery WR (2001) Cavefish as a model system in evolutionary developmental biology. Dev Biol 231:1–12. doi:10.1006/dbio.2000.0121

Jeffery WR (2008) Emerging model systems in evo-devo: cavefish and microevolution of development. Evol Dev 10:265–272

Katz B, Minke B (2009) *Drosophila* photoreceptors and signaling mechanisms. Front Cell Neurosci 3:1–18. doi:10.3389/neuro.03.002.2009

Keightley PD, Trivedi U, Thomson M, Oliver F, Kumar S, Blaxter ML (2009) Analysis of the genome sequences of three *Drosophila* melanogaster spontaneous mutation accumulation lines. Genome Res 19:1195–1201. doi:10.1101/gr.091231.109

Khan AI, Dinh DM, Schneider D, Lenski RE, Cooper TF (2011) Negative epistasis between beneficial mutations in an evolving bacterial population. Science 332:1193–1196. doi:10.1126/science.1203801

Kim EB, Fang X, Fushan AA, Huang Z, Lobanov AV, Han L, Marino SM, Sun X, Turanov AA, Yang P, Yim SH, Zhao X, Kasaikina MV, Stoletzki N, Peng C, Polak P, Xiong Z, Kiezun A, Zhu Y, Chen Y, Kryukov GV, Zhang Q, Peshkin L, Yang L, Bronson RT, Buffenstein R, Wang B, Han C, Li Q, Chen L, Zhao W, Sunyaev SR, Park TJ, Zhang G, Wang J, Gladyshev VN

(2011) Genome sequencing reveals insights into physiology and longevity of the naked mole rat. Nature 479:223–227

Kishimoto T, Iijima L, Tatsumi M, Ono N, Oyake A, Hashimoto T, Matsuo M, Okubo M, Suzuki S, Mori K, Kashiwagi A, Furusawa C, Ying B-W, Yomo T (2010) Transition from positive to neutral in mutation fixation along with continuing rising fitness in thermal adaptive evolution. PLoS Genet 6:e1001164

Lazard D, Zupko K, Poria Y, Net P, Lazarovits J, Horn S, Khen M, Lancet D (1991) Odorant signal termination by olfactory UDP glucuronosyl transferase. Nature 349:790–793. doi:10.1038/349790a0

Lencz T, Lambert C, DeRosse P, Burdick KE, Morgan TV, Kane JM, Kucherlapati R, Malhotra AK (2007) Runs of homozygosity reveal highly penetrant recessive loci in schizophrenia. Proc Natl Acad Sci USA 104:19942–19947. doi:10.1073/pnas.0710021104

Luque T, O'Reilly DR (2002) Functional and phylogenetic analyses of a putative *Drosophila* melanogaster UDP-glucosyltransferase gene. Insect Biochem Mol Biol 32:1597–1604. doi:10.1016/s0965-1748(02)00080-2

Mackay TFC, Richards S, Stone EA, Barbadilla A, Ayroles JF, Zhu D, Casillas S, Han Y, Magwire MM, Cridland JM, Richardson MF, Anholt RRH, Barron M, Bess C, Blankenburg KP, Carbone MA, Castellano D, Chaboub L, Duncan L, Harris Z, Javaid M, Jayaseelan JC, Jhangiani SN, Jordan KW, Lara F, Lawrence F, Lee SL, Librado P, Linheiro RS, Lyman RF, Mackey AJ, Munidasa M, Muzny DM, Nazareth L, Newsham I, Perales L, Pu L-L, Qu C, Ramia M, Reid JG, Rollmann SM, Rozas J, Saada N, Turlapati L, Worley KC, Wu Y-Q, Yamamoto A, Zhu Y, Bergman CM, Thornton KR, Mittelman D, Gibbs RA (2012) The *Drosophila melanogaster* genetic reference panel. Nature 482:173–178

McQuillan R, Leutenegger A-L, Abdel-Rahman R, Franklin CS, Pericic M, Barac-Lauc L, Smolej-Narancic N, Janicijevic B, Polasek O, Tenesa A, MacLeod AK, Farrington SM, Rudan P, Hayward C, Vitart V, Rudan I, Wild SH, Dunlop MG, Wright AF, Campbell H, Wilson JF (2008) Runs of homozygosity in European populations. Am J Hum Genet 83:359–372. doi:10.1016/j.ajhg.2008.08.007

Metzker ML (2010) Sequencing technologies – the next generation. Nat Rev Genet 11:31–46. doi:10.1038/nrg2626

Mori S (1986) Changes of characters of *Drosophila melanogaster* brought about during the life in constant darkness and considerations on the processes through which theses changes were induced. Zoolog Sci 3:945–957

Mori S, Yanagishima S (1959) Variations of *Drosophila* in relation to its environment VII Does *Drosophila* change its characters during dark life? Jpn J Genet 34(1):151–161

Ng J, Luo L (2004) Rho GTPases regulate axon growth through convergent and divergent signaling pathways. Neuron 44:779–793. doi:10.1016/j.neuron.2004.11.014

Partridge L, Fowler K (1990) Non-mating costs of exposure to males in female *Drosophila* melanogaster. J Insect Physiol 36:419–425. doi:10.1016/0022-1910(90)90059-o

Payne F (1911) *Drosophila* ampelophila Loew bred in the dark for sixty-nine generations. Biol Bull 21:297–301

Protas M, Conrad M, Gross JB, Tabin C, Borowsky R (2007) Regressive evolution in the Mexican cave tetra, *Astyanax mexicanus*. Curr Biol 17:452–454. doi:10.1016/j.cub.2007.01.051

Ravi Ram K, Wolfner MF (2007) Seminal influences: *Drosophila* Acps and the molecular interplay between males and females during reproduction. Integr Comp Biol 47:427–445. doi:10.1093/icb/icm046

Rogge RD, Karlovich CA, Banerjee U (1991) Genetic dissection of a neurodevelopmental pathway: son of sevenless functions downstream of the sevenless and EGF receptor tyrosine kinases. Cell 64:39–48. doi:10.1016/0092-8674(91)90207-f

Schmidt A, Hall A (2002) Guanine nucleotide exchange factors for Rho GTPases: turning on the switch. Genes Dev 16:1587–1609. doi:10.1101/gad.1003302

Wang T, Montell C (2007) Phototransduction and retinal degeneration in *Drosophila*. Pflugers Arch 454:821–847. doi:10.1007/s00424-007-0251-1

Wang Q, Hasan G, Pikielny CW (1999) Preferential expression of biotransformation enzymes in the olfactory organs of *Drosophila melanogaster*, the antennae. J Biol Chem 274:10309–10315. doi:10.1074/jbc.274.15.10309

Weinert BT, Timiras PS (2003) Invited review: theories of aging. J Appl Physiol 95:1706–1716. doi:10.1152/japplphysiol.00288.2003

Williams JA, Su HS, Bernards A, Field J, Sehgal A (2001) A circadian output in *Drosophila* mediated by Neurofibromatosis-1 and Ras/MAPK. Science 293:2251–2256

Yamamoto Y, Stock DW, Jeffery WR (2004) Hedgehog signalling controls eye degeneration in blind cavefish. Nature 431:844–847. doi:10.1038/nature02864

Yamamoto Y, Byerly MS, Jackman WR, Jeffery WR (2009) Pleiotropic functions of embryonic sonic hedgehog expression link jaw and taste bud amplification with eye loss during cavefish evolution. Dev Biol 330:200–211. doi:10.1016/j.ydbio.2009.03.003

Yoshizawa M, Goricki S, Soares D, Jeffery WR (2010) Evolution of a behavioral shift mediated by superficial neuromasts helps cavefish find food in darkness. Curr Biol 20:1631–1636. doi:10.1016/j.cub.2010.07.017

Yuan Q, Xiang Y, Yan Z, Han C, Jan LY, Jan YN (2011) Light-induced structural and functional plasticity in *Drosophila* larval visual system. Science 333:1458–1462

Zhou D, Udpa N, Gersten M, Visk DW, Bashir A, Xue J, Frazer KA, Posakony JW, Subramaniam S, Bafna V, Haddad GG (2011) Experimental selection of hypoxia-tolerant *Drosophila melanogaster*. Proc Natl Acad Sci USA 108:2349–2354. doi:10.1073/pnas.1010643108

Zhu J, DeLuca HF (2012) Vitamin D 25-hydroxylase – four decades of searching, are we there yet? Arch Biochem Biophys. doi:10.1016/j.abb.2012.01.013

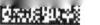